Genetically Modified Planet

Genetically Modified Planet

Environmental Impacts of Genetically Engineered Plants

C. Neal Stewart, Jr.

2004

OXFORD
UNIVERSITY PRESS

Oxford New York
Auckland Bangkok Buenos Aires Cape Town Chennai
Dar es Salaam Delhi Hong Kong Istanbul Karachi Kolkata
Kuala Lumpur Madrid Melbourne Mexico City Mumbai Nairobi
São Paulo Shanghai Taipei Tokyo Toronto

Published by Oxford University Press, Inc.
198 Madison Avenue, New York, New York 10016
www.oup.com

Library of Congress Cataloging-in-Publication Data
Stewart, C. Neal.
Genetically modified planet : environmental impacts of genetically
engineered plants / C. Neal Stewart, Jr.
p. cm.
Includes bibliographical references
ISBN 0-19-515745-1
1. Transgenic plants. 2. Plant genetic engineering. 3. Transgenic
plants—Ecology. 4. Plant genetic engineering—Environmental aspects.
I. Title.
SB123.57 .S74 2004
631.5'233—dc22 2003021371

9 8 7 6 5 4 3 2 1

Printed in the United States of America
on acid-free paper

To my wife, Susan Stewart, for her love, support, kindness, and patience.

Preface

This book is the natural outgrowth of several previous writing projects, including book chapters, review papers, and journal articles, public and professional presentations, and an undergraduate class I taught on the risks and benefits of biotechnology. In the midst of my course, I realized how much dubious information about biotechnology the typical freshman had accumulated in his or her brief lifetime. It became apparent that a book written for a general audience on the science of the environmental impacts of genetically modified plants was needed. In the process of writing this book, I have learned much about the science and even more (I hope) about communicating the subject to nonscientists. I've learned firsthand that it truly is easier to write to my peers using shop jargon than to effectively communicate to the people who are, perhaps, most concerned about the ramifications of science and technology. And that truly is the purpose of this book: to communicate and demystify biotechnology to the degree that the reader can go beyond newspaper headlines to critically examine the issues. There is power in the understanding of relative risks and benefits of any technology. There is certainly value in better understanding agriculture—to go beyond the supermarket to visit the source of where our food comes from—and examining the role of genetics in farming. If this book accomplishes nothing else, I'll be satisfied to know that people have a better grasp of how agriculture really works as a result of reading it.

Many people have enabled the writing of this book. The members of my lab crew always amaze me with their hard work and insights. Without these talented people, I would have had neither the knowledge nor the time to write such a book. Specifically, Matt Halfhill has performed much of the actual ecological biosafety research in my lab, with the aid of several other scientists. I also want to thank Matt for his critical reading of the manuscript. Thanks to my principal collaborators in this field, John All, Paul Raymer, and Suzanne Warwick, who have been instrumental in helping me see the big picture of agriculture, along with my postdoctoral mentor, Wayne Parrott, who encouraged my entrance into this intriguing area of research. Likewise, I have benefited from stimulating interactions from other collaborators in the field, including Angharad Gatehouse, Angelika Hilbeck, Guy Poppy, and Tanja Schuler. The area of biotechnological biosafety has many sides, though, and conversations with people such as Tom Nickson, Detlef Bartsch, Debby Sheely, Jeremy Sweet, Fred Gould, Greg Warren, Anna Hope, Mark Sears, and C.S. Prakash have also definitely shaped my opinions.

Several people were instrumental in the shaping and wordsmithing of this book. I am indebted to Kirk Jensen, former executive editor at Oxford University Press, for his guidance and editing, and Sarah Wheaton for her help on the book proposal. I am grateful to John Rauschenberg, Heather Hartman, and the Oxford staff. Sandy Kitts at the University of Tennessee was kind enough to render several illustrations. Amy Yancey Jenkins designed the cover, and Susan Stewart, Reggie Millwood, Mentewab Ayalew, and Nathan Stewart provided drawings for illustrations. In addition, Reggie runs my lab day-to-day, and his diligence and trustworthiness have been critical in my time of sequestration on this project. I am especially thankful to Susan Stewart, who played a vital role in the final revisions of several chapters. The writing of this book (time, support, and information) was facilitated by a research leave provided by the University of North Carolina at Greensboro; the USDA Biotechnology Risk Assessment Research Grants Program; the support of the Racheff Chair of Excellence endowment; and the support of staff and administrators of the University of Tennessee Department of Plant Sciences and the University of Tennessee Institute of Agriculture. Finally, I am indebted to my parents, Charles Stewart, Sr., and Jane Stewart, as well as to my good friend Richard Fredrickson, who were quite instrumental in helping foster in me a love of nature, which is the reason I became a scientist in the first place.

All opinions and errors are my own and do not necessarily reflect the opinions of the University of Tennessee or any of the above people or institutions. To God be the glory.

Contents

Genetically Modified Planet

1

Introduction

Catastrophic Calamities and Clucking Cacophonies

A recent conversation with a major beneficiary of agricultural biotechnology revealed the depth of the American commitment to planting genetically modified (GM) crops. The proponent, a farmer in western Tennessee, explained why GM varieties are popular. "In the field, enemy number one is weeds. Roundup Ready soybeans help me control weeds easier with fewer chemicals." The soybeans he planted have an additional bacterial gene that allows them to thrive when fields are sprayed with the chemical glyphosate, the active ingredient in the herbicide Roundup. Only the weeds die. The farmer continued to explain why Roundup Ready soybeans top the list of GM crops grown on American soil: "If weeds can be controlled early in the season, the soybeans form a canopy that shades out weeds. Before these GM soybeans became available I was having to use several herbicides and tillage to keep ahead of weeds." GM soybeans have significantly facilitated reduced tillage practices, and low-till and no-till farming result in less soil erosion. When asked about real-life benefits, the farmer stated, "The bottom line for me is that I can make more money because I spend less for herbicide applications. I get good yields with these beans, plus my soil stays on the farm instead of going down the [Mississippi] River."

Biotechnological crops are indeed popular with farmers. In the United States in 2003, GM crops composed approximately one-third of the corn acreage and three-quarters of areas planted in cotton and soybeans. Most of the corn and cotton crops had a single gene addition that enabled them to produce their own insecticidal proteins; the plants now kill insect pests without pesticides. Insect-resistant biotech cotton has eliminated the need to spray millions of gallons of synthetic chemical insecticides, a huge, direct environmental benefit.

Farmers in several other countries have also embraced GM crops. In Canada, more than 70% of the canola crop consists of herbicide-tolerant varieties. Argentina and China have extensive acreage planted in transgenic crops.

Although GM plants are not very controversial in North and South America and China, they are controversial in other places, particularly in Europe. A number of vocal activists believe that GM crops may be harmful to the environment. They believe that biotechnology is inherently dangerous and that the new plants have not been sufficiently tested to assure environmental safety. One vehement opponent is Dr. Mae-Wan Ho, a reader at the Open University in England. She has declared, "Genetic-engineering biotechnology . . . will spell the end of humanity as we know it and of the world at large."[1] This statement represents an extreme viewpoint. While most anti-GM activists don't subscribe to a doomsday prediction, there are plenty of people who share Ho's general misgivings about the safety of agricultural biotechnology. But science can answer the biosafety questions.

For the sake of argument, let's say that Ho is correct and GM plants will have a significant negative impact on the world. Ecological disaster would certainly emerge first in the United States, where GM crops are most popular. In America, millions of acres have been planted with engineered crops (tables 1.1 and 1.2). Since the first GM plants were grown in the field experimentally in 1988, an estimated 38 trillion GM plants have been grown in American soil; the plants were modified for dozens of traits with hundreds of different transgenes (table 1.3). More than 99.9% have been grown under deregulated conditions, indicating that American regulatory authorities are convinced that GM plants are safe. Ecological disaster has, thus far, remained at bay.

At one end of the spectrum, proponents of biotechnology argue that GM plants are as safe as conventionally bred crops. At the other end, Ho and a montage of activists decry danger. Who's right?

There are real and perceived environmental risks associated with GM plants. Unfortunately, the perceived risks have often been greatly elevated in their stature. Although some of these erstwhile dangers might have nuggets of truth, others are nothing more than dreamy arguments. Nonetheless, I will address them all, in the context of weighing risks against

Table 1.1. Typical numbers of plants grown per acre in the United States

Crop	Plants per acre
Corn	24,000
Cotton	40,000
Soybean	120,000
Canola	350,000
Potato	40,000

These figures are drawn from consultations with agronomists.

benefits and the dangers of accepting a form of technology against hazards of rejecting it. Although benefits of any methodology or invention might be enormous, if it is inherently and "fatally" hazardous, acceptance of the technology is precarious at best. A prime example is nuclear energy, which promised to produce abundant, cheap energy. It did not take many "small" nuclear disasters for the public to decry future construction of nuclear reactors. Yet we regularly employ risky technology. The difference between what we take for granted and what we worry about often boils down to familiarity and perceived necessity. There are multiple risks of crashing and bodily injury in automobile travel; however, automobile travel is generally considered safe. There are safety issues

Table 1.2. Estimates of acres (in millions) of GM crops in the United States

	Year								
Crop	pre-1996	1996	1997	1998	1999	2000	2001	2002	2003
Corn				19.6	28.3	19.9	19.8	24.5	24.5
Cotton				5.8	7	9.3	11.1	10.1	10.1
Soybean				27	35	40.2	51.3	54.0	54.0
Canola				0.4	0.5	0.9	0.8	0.8	0.7
Potato				0.06	0.06				
All (mainly soy and corn) before 1997	1	6	18						

The USDA National Agricultural Statistics Service (NASS) and Biotechnology Industry Organization (BIO) make estimates of the yearly acreages of GM crops grown in the United States.

Table 1.3. Numbers of GM plants grown in the United States (in millions)

Crop	Year									Totals
	Pre-1996	1996	1997	1998	1999	2000	2001	2002	2003	
Corn				470,400	679,200	477,600	475,200	588,000	588,000	3,278,400
Cotton				232,000	280,000	372,000	444,000	404,000	404,000	2,136,000
Soybean				3,240,000	4,200,000	4,824,000	6,156,000	6,480,000	6,480,000	31,380,000
Canola				140,000	175,000	315,000	280,000	280,000	245,000	1,435,000
Potato				2400	2400	0	0	0	0	4800
Totals	80,000	480,000	1,440,000	4,294,800	5,511,600	5,848,600	7,092,700	7,489,500	7,489,500	38,234,200

Numbers were derived by multiplying the appropriate figures from tables 1.1 and 1.2.

associated with eating: food poisoning, allergies, and diseases associated with poor choices. Despite risks of vomiting, anaphylactic shock, and growing morbidly fat, though, most people would agree that food consumption is safe. Risk management for these and other familiar technologies is approached with a degree of scientific objectivity. It is my hope that scientists and the general public will likewise attain an objective analysis for the ecology of GM plants. One goal of this book is to demystify agricultural biotechnology and place it in a more familiar context.

I will not cover GM food safety issues in this book. Testing for toxic compounds and allergenicity is more straightforward than ecological study. Animal feeding experiments can pinpoint the acute toxicity of a novel food, and there are numerous predictors for determining whether a modified consumable will cause food allergies.

Ecological biosafety is a more complex science. Nature is a big place, and ecological interactions are far from being completely understood. In many ways, the precise introduction of single genes into plants and the analysis of their new ecological interactions can provide an excellent toolkit to better understand fundamental biology. But here we are interested in how research increases our understanding about the biosafety of GM plants.

Certain risks appear repeatedly in the scientific literature and popular press, each presenting drastic outcomes. I will explore the likelihood of environmental calamity for each of several issues. One concern is that insect pests might develop resistance to a transgenic, plant-produced pesticide and become more difficult to control. A second is that gene flow might occur between GM crops and weeds to produce superweeds of extraordinary competitive ability. A third risk is that viral transgenes in plants could recombine with existing viruses to produce superviruses.

In recent years, these and other fears have been spun into hot news stories. One of the better-known reports focused on GM corn and the monarch butterfly; early indications of a potential side effect were translated into a dramatic ecological tragedy for public consumption. In the monarch butterfly case and in other instances, actual empirical knowledge never arrived in newspapers and TV news. We want to be able to untangle the science from publicity.

The prevalent belief among skeptics is that once GM plants are released into the environment, the plants or genes can never be recalled. Once the banana is out of the peel, it is impossible to put it back. Is the cultivation of 38 trillion GM plants a huge, uncontrolled American experiment with potentially devastating consequences? Are the prognostications of dire ecological consequences reasonable warnings or Luddite ravings? Are GM plants simply an extension of conventional agriculture, or are we irreversibly treading on Mother Nature?

My study of GM plants began in 1993, and my interest was generated by the scientific possibility that certain plant–transgene combinations might carry significant and unique environmental risks. Since these beginnings I have changed from being a biosafety skeptic to a cautious optimist and a proponent of biotechnology. Still, I remain a strict advocate and practitioner of biosafety research, in favor of appropriate regulatory frameworks for the sake of environmental protection. I now believe that GM technology will not destroy the earth, but may well make the world a safer, cleaner, and greener place in which to live. I am a practicing plant ecologist and molecular biologist—an oddball in the world of biology. It is not common for a single scientist to both produce GM plants and study their ecology, as the type and scale of the research done in these two areas are usually quite different. Ecologists study organisms interacting with other organisms and with their environments. The spatial scale of interest typically spans somewhere between several square meters to square kilometers and beyond. The ecological timeline is usually months, years, decades, or longer. On the other hand, molecular biology is performed at the submicroscopic to whole-organism level, and experiments generally run from minutes to days. The two areas are rarely married, but to understand fully the interactions between a GM plant and its environment in the real world, along with the role of GM plants, a dialogue between molecular biology and ecology is extremely helpful and perhaps fundamentally essential. Molecular biology and ecology intersect at the synthetic field of research known as biotechnology risk assessment research, the ecology of transgenic plants.

My interest in the biosafety of GM plants was spurred by a study published in 1993 that described experiments in which herbicide-tolerant GM plants were grown in the field and found to be completely ecologically benign.[2] These early experiments intrigued me. The trait the transgene provided, herbicide tolerance, would not be expected to provide an increase in fitness or competition to its host—and it didn't. But the responsible British group used a very appropriate plant: canola, an oilseed crop in the mustard family that has several wild relatives with which it can interbreed.

In 1994, I began studying what I believed to be a worst-case scenario for a simple GM crop (containing one or two novel transgenes), which is described in more depth in chapter 4. Like the 1993 study, the research used canola along with its wild relatives, but different transgenes. I thought that a worst-case transgene would be one that could be affected by natural selection outside of agriculture, a *Bacillus thuringiensis* (Bt) transgene that encodes a caterpillar-killing protein. The hypothesis tested was that harboring a Bt gene could potentially make canola a weed, or confer weedy or invasive tendencies to help its survival outside an agricultural field. In addition, the Bt transgene could be transferred through hybridization and introgression from the crop to one of its weedy wild relatives. Would the transgene make the weed even weedier? It seemed to me that the wrong transgene (one that could confer fitness-enhancing traits) in the wrong crop (one that could transfer genes to weeds) could result in an ecological disaster. In such a case, the interaction between the transgene's unique and profound effect, coupled with the ecology of the crop and weed, would possibly result in an environmental effect greater than the sum of its parts. The wrong combination could create a new kudzu, some superweed of the highest order, even worse than that exotic legume imported into the southeastern United States for erosion control that now swallows homesteads with a single-season gulp.

In the following chapters, I examine the ecological risks of several crop–transgene combinations from a scientific perspective. Of interest is how the media and others interpret scientific research as these entities influence popular opinion. We'll take a look at views postulated by those who are opposed to GM technology in general and try to reach some middle ground, at least as far as the science will allow us to go. In the end, we'll see where scientific knowledge ends and speculation begins. Into the foreseeable future scientists will continue to perform experiments pinpointing risks and benefits in order to increase our knowledge about the roles of GM plants in agriculture. This research, no doubt, will continue as long as there are GM crops.

When the risks and benefits are diverse, how do we determine the advantages and acceptability of a technology, such as biotechnology? I think the safest path is to examine several specific cases where extensive scientific knowledge is available to extrapolate to bridges we're bound to cross in the future. And we will look to the future of GM plants. Plants that will be used to clean up the environment and monitor toxins will help us address problems we have not been able to solve in any other way.

The book begins and ends with some philosophical musings intended to place biological research and developments, and, in fact, agriculture, into a framework of the existing natural world. This world is highly modified by, among other things, agriculture and its blatantly unnatural components. Will modifying agriculture with biotechnology alter nature even more than it already has? Or will biotechnology actually make agriculture more harmonic with nature? We can learn a great deal about nature, agriculture, and genetic engineering by examining existing crops and weeds. The context for genetic alterations is in the farmer's field.

2

Crops and Weeds

It's Hard to Be a Wild Thing When You're Domesticated

On the Beach au Naturale

What is natural? It is an important question for interpreting biotechnology. As I sat au naturale on a deserted beach in Australia, I pondered that question. Certainly, in that setting, it appeared that everything I was seeing was natural: sea, sand, rocks, birds, and a horizon that consisted of commingled water, sky, and clouds. On other days, I've walked trails in old-growth forests observing that this, too, seemed completely natural. In each setting there was only one glaringly unnatural thing: me! There I was, a human being, sporting clothes (sometimes), a wristwatch, money, and keys—all of which linked me to the rest of the seemingly unnatural, man-made world. I, a member of the ultimate invasive species, was a foreign trespasser in otherwise pristine environments.

There is a contrasting view of nature that is all-inclusive. Here humans are as organic as the next species and part and parcel of nature itself. All organisms adapt to their environment, seeking to maximize their resource requisition and their fitness. Humans are no different from the rest of life, and while they may have modified their environment with seemingly artificial things, they are still members of nature, along with every other organism and its environmental modifications. In this case, automobiles, airplanes, asphalt, and office buildings are all part of nature, albeit a human-modified nature, which seems to be the natural thing for humans to do! In either view, mosquitoes, poison ivy, and rabies are also natural.

I think that most people's opinion of "nature" or "natural" would line up more closely with the first viewpoint than with the second "humans-as-natural-invaders" scenario. The latter is more logical, but the first is more emotionally satisfying. In reality, it is quite difficult to differentiate between the natural and unnatural. Is it natural for a chimp to use a stick to flush a dinner of ants from a hole in the ground? Is it natural for me to catch a fish dinner from a stream using another type of tool, a rod and reel? What is the difference between chimps and humans using tools to modify their environments and acquire resources? While the fishing experience might make me feel quite connected to nature, am I really invading a natural ecosystem using a tool, quite similar to the chimp, to catch food? Does the fisherman even have a natural environment? If so, where do we go to find it? Is it natural for me to be sitting in my office and typing on a computer? If I must be typing, is that my natural environment? Many of my colleagues might assume that it is, but if I had to be typing, I'd rather be typing on a beach in Australia.

This little exercise in futility is simply to underscore the point that we really can't precisely define what is natural. Many of the things that we assume are completely natural can't logically be declared as such. Agriculture and food lie at the epicenter of claims and confusion about all things natural. A case in point is so-called natural or organic food. Nearly all of it comes from crops that are overbred and highly domesticated. "Natural" wheat crackers come from a polyploid species that is a combination of at least three other species. Wheat is the product of millennia of human intervention and innovation. The crop is grown in monocultures designed to maximize the yield of one certain plant seed at the expense of every other plant, animal, and microbe—a low-biodiversity environment. After the grain is harvested with a machine, it is shipped in a truck across the country and loaded into a mechanical crusher. The crushed wheat is then mixed with other unnatural ingredients and chemicals, cooked in a man-made oven, and cut into uniform geometrical shapes. The so-called natural wheat crackers are packaged in petroleum-derived plastic and then secondarily packaged in a box made from dead and highly treated trees. Finally, the boxes are shipped to a grocer in a truck fueled by more petroleum and stocked on shelves in a place that, once again, is designed to deliver food to humans only. The crackers are often eaten with "natural" cheese that is produced using one of the earliest-developed processes of biotechnology involving the manipulation of enzymes.

So what makes such crackers natural? Is it because the only purified chemical they may contain is sodium chloride? Or is it because the wheat was grown on a farm that used no GM crops, synthetic fertilizer, or pesticides? If the latter were the case and the farm were "organic," crops may have been sprayed with a naturally occurring bacterial pesticide, such as *Bacillus thuringiensis* (a relative of the anthrax-causing bacterium; see figure 2.1), or otherwise contaminated with fungi that produce known neurotoxins.

The fact remains that the terms "natural" and "nature" are not terms we use to describe an objective reality, but are based largely on human emotion and aesthetics. Nature is not precisely definable. That is not to say that our perceptions of nature are without value—the increasingly rare environments that have not been overtly modified by humans should be preserved. There is emotional, social, spiritual, and even economic value in their conservation. But even these so-called natural environments contain blatantly unnatural components, components that were not present in recent history. Just behind my pristine, deserted beach in Australia was a semi-managed strip of lawn and tangled bank that contained *Vinca major*, large periwinkle, a popular landscaping groundcover (figure 2.2). Also present were populations of *Raphanus raphanistrum*, the wild radish, a particularly invasive agricultural weed (figure 2.3). In addition to these two species, there were a number of other escaped ornamental plants and exotic grasses. Part of the strip was intermittently mowed, but it was

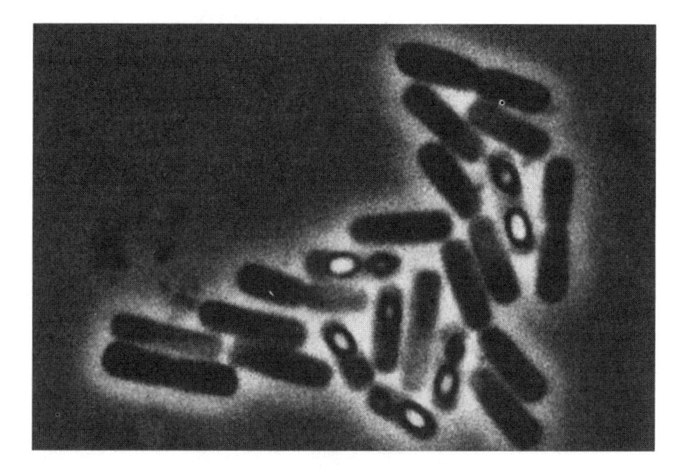

Figure 2.1. *Bacillus thuringiensis* bacteria and insecticidal crystals are used as insect control agents.

Figure 2.2. A natural-looking beach in Victoria, Australia. Note that the vegetation to the left of the central path contains horticultural and weedy plants that are not native to the continent. The bottom left contains a large patch of wild radish (*Raphanus raphanistrum*).

Figure 2.3. A prolific population of Australian wild radish on the beach.

otherwise unmanaged, and it seemed that no one had purposefully planted or maintained the bank. To most people, the plants would appear to be part of nature. To the botanical eye, though, they could be immediately recognized as exotic invaders.

Most people would enjoy seeing the small stands of wild radish with its light yellow flowers attracting a number of butterflies. Those same people would react less favorably to a monoculture of it by the beach, severely decreasing the biodiversity of other kinds of plants that could have been growing there. So exactly what are our baseline and criteria for defining natural or invasive species? Does it matter how long ago wild radish invaded the Australian coastline? Let's imagine it had been there for 20,000, 300, 20, or only 2 years. Where do we draw the line between native or natural and exotic or unnatural? These exercises are not just academic calisthenics. If the wild radish population had been there for 20,000 or even 300 years, we might rightly assume that the species had reached equilibrium, and radish would likely not displace other plants and otherwise alter stable ecosystems in the future. Twenty years may not have been long enough for population stasis to be reached. As for a plant species that suddenly appeared within the past year, it is anyone's guess what it will do. It might die out the following year, or it could become the new Australian kudzu. We want to avoid anything that could become the latter.

In actuality, wild radish has probably been along this particular strand of beach for about 15–20 years, and it is likely to continue its expansion, both in number of patches and size of patches. We can make this prediction based on its ecology. It is a nasty weed in grain crops in Australia and all other grain-growing continents; one author has called it "the nemesis of man's grain crops."[1] Most ecologists, weed scientists, and agronomists, as well as farmers and naturalists, would agree that wild radish is a problem for Australia that will only get worse. It has a prolonged vegetative phase in the spring and then starts flowering and setting seeds, continuing to do so for an extended period of time. In contrast, crops tend to flower and set seed in a linear fashion and in a short period of time. Wild radish seeds are in fruits called siliques that do not easily release their propagules. In my lab and other labs that work with this weed, researchers have invented various and sundry apparatuses to coax the seeds out of the wild radish siliques without crushing the seeds, or without driving the seed-extracting operator into a frenzy. That the seeds stay in the siliques a long time endows the species with a mechanism that staggers seed germination, called seed banking. Most crop plants do not have this ability; seed dormancy is a trait of many weeds and invasive plants that allows them the flexibility to wait and germinate under optimum conditions.

What Makes a Weed a Weed and a Crop a Crop?

As evidenced in the above example, a plethora of traits tend to separate crops from weeds and invasive plants (table 2.1). Wild radish's ability to hold seeds into fruits and then stagger seed germination are two examples of weed traits that crops generally don't have. These are two of the more important traits that make a plant weedy. More of these traits are listed in table 2.1.

It should be noted from the outset that most crop plants are lousy weeds. They certainly don't compete well with native plants outside of agriculture or invade natural ecosystems. In fact, many crop plants don't compete well against weeds inside agriculture either, which is why farmers go to great lengths to control weeds. Control of weeds is the primary pest-associated job in farming, as they cause over one billion dollars of annual crop losses in the United States alone. The reason crops are not very effective weeds is that they have been overbred and selected for traits that are good exclusively for farming, which are often opposite to those required for a successful weed. A crop breeder wants to turn out a completely domesticated plant that does one thing well—produce readily harvestable, high seed production all at once. A good crop plant will not

Table 2.1. A list of weediness traits

1. Ability to produce seeds early in life cycle
2. Profuse reproduction by seeds and/or vegetative structures
3. Seed dormancy and asynchronous germination; seed banks
4. Adaptations to coexist and be spread with crop seeds
5. Production of allelochemicals, biological toxins that suppress the growth of other plants
6. Prickles, spines, or thorns that can cause physical injury and repel animals or that can enable seed dispersal
7. Ability to parasitize other plants
8. Large food reserves by underground structures
9. Survival and seed production under adverse environmental conditions
10. High photosynthetic capacity and growth rates
11. Fibrous root systems that can effectively forage for nutrients

After Baker (1965).[2]

only churn out high-quality seeds, but should also keep its seeds in the pod until harvested. Spilling seeds on the ground is known as "shattering," a poor trait for a crop variety. The breeder also doesn't want the crop plant to waste its resources by producing allelochemicals, spines, or profuse roots and rhizomes (with some good exceptions like carrot). But you'll notice here that wild radish doesn't shatter either. A successful weed need not have all the traits listed in table 2.1, but it certainly takes the right combination of traits to successfully outcompete crops.

Humans have been genetically modifying plants for millennia for the main purpose of increasing domestication—turning wild plants into tame ones. Croplike traits have been selected in several food species for at least 10,000 years. While prototypical crop breeders and early farmers did not understand genetics or realize what they were doing at the molecular level, they in fact gradually modified wild species into plants that were more manageable and predictable. They created entities we recognize today as crops. Humans define which plants are crops and which plants are weeds. Of course, all the basic genetic building blocks have always been natural, and still are; DNA and proteins are as natural as it gets. But the outcome of the plodding genetic modification is that while some plants are now domesticated, the wild relatives are also with us still. Wild carrot is called Queen Anne's lace and is known for its delicate, lacy, white, decorative flower. It also has a white root. Domesticated carrot has an orange, nutritious, and tasty root. Wild carrot and crop carrot are interfertile and, in fact, in the same species, *Daucus carota* (figure 2.4). Crops were created by a series of breeding and selection episodes in which off-types were discarded and advantageous types were kept. Multiple genes and traits were retained simultaneously. In fact, no crops, as we know them now, existed 12,000 years ago. They were all fabricated by humans and for humans. Domesticated crops share the basic and natural molecular makeup that all other organisms possess: nucleic acids, proteins, carbohydrates, and lipids. The distinct entities that we recognize as crops, though, are essentially human inventions. See figures 2.4 and 2.5 for the agricultural development of carrot and *Brassica oleracea* vegetables. Nature would have certainly never stooped so low as to give us brussels sprouts and kohlrabi.

Although there is very little truly natural about agriculture, it is difficult to find anything harmful arising from human intervention in plant genetics, whether it be traditional sexual crosses, mutagenesis breeding, or modern biotechnology. Humans have not converted safe plants into unsafe plants. In fact, quite the opposite has been the rule. We have bred

Wild Carrot Domesticated Carrot

Figure 2.4. Carrot domestication. Wild carrot, *Daucus carota*, is also known as Queen Anne's lace. Orange carrots appeared in Holland in the 1700s. The phenotypic differences between wild and domesticated carrot are significant. (Photos by Edward Chester and Photos.com.)

plants that no longer produce the toxins that are so common in the natural vegetation of the world, toxins that protect wild plants from being eaten. The most tragic outcome of human intervention in the plant world has not been genetic but ecological: introducing plants in new places that were not part of their historical distributions. Ironically, the excuse has most often been under the auspices of ornamental landscape improvement and not food production. The genetic differences we typically observe in domesticated plants are neutral or crippling for the crop; plants that are chosen for domestication are not invasive. Crops do not make good weeds or even good "natural" plants. It has been said that when humans disappear from earth, corn will follow the very next year. *Zea mays* ssp. *mays* cannot survive without us.

The following points and illustrations might help us better understand the differences between weeds and crops. Weeds and invasive plants have a suite of traits that work together to make them weeds. Likewise, crops have a suite of traits that make them domesticated. A typical plant (crop

Figure 2.5. The evolution of *Brassica oleracea* crops. There are numerous forms within this species. (Photos by Paul Williams.)

or weed) will have between 15,000 and 100,000 genes. The exact number is species specific and, in most cases, still unknown. The differences we observe between plant species and types are the result of different combinations of genes. A single unique gene does not make a plant either a weed or a crop. Many critics of biotechnology have drawn the comparison between genetically modified plants and invasive species. They believe that introducing a single novel gene into the environment through a crop plant is analogous to introducing a foreign species, and they don't want another kudzu to be created by accident (figure 2.6). Kudzu was introduced into the southern United States from southeastern Asia in the nineteenth century as an ornamental plant. It was then subsequently widely planted to control erosion along roadsides and other steep embankments. It is now a feral weed on millions of acres in the southeastern United States, where it continues to spread. Is the introduction of an exotic species in the form of 50,000 genes packaged together in one plant (e.g., kudzu) the same thing as introducing a single novel gene in a well-characterized crop plant? The answer is a resounding no.

Singing the Single-Gene Blues

But what if the transgene moves from a genetically engineered crop to a weed that already has a lot of those weediness traits listed in table 2.1? This has been an important research topic in my lab because the answer is pivotal for the future of biotechnology in agriculture. Although we are still in the midst of the research and don't have the definitive and con-

Figure 2.6. Kudzu covering trees in North Carolina. In the southeastern United States, kudzu, an exotic weed, can rapidly coat vegetation, houses, and slow-moving pets.

clusive answer yet, there are some biologically sound arguments that suggest that a single transgene or even several transgenes will not transform a crop into a weed or change an ordinary weed into a superweed. The following illustration might help explain why.

Think of the number of words in this book. It is somewhere around 75,000 words; the number is in the range of the number of genes in a typical plant. If we were to introduce another word, let's say an obscene word or a very technical and scientific-sounding word in the middle of the book, what would happen? There are two possibilities that are quite predictable. First, the sentence in which the word landed might not make very much sense. One could view it as a typo. Second, it is conceivable that the single word insertion might even change the meaning of a sentence or even perhaps a paragraph. But it would not change this GM plant book into a steamy romance novel, or a cookbook, or a biography on the life and times of the musician Ben Folds, even if it were a dirty word. The single word insertion would result in a very, very slightly altered GM plant book. The change would be minuscule in the grand scheme of the book.

A similar paradigm can be expected when a single gene is inserted into a plant genome, which is exactly what genetic engineering does. Using the current technologies, we don't know into which sentence, paragraph, or even chapter the word will go, but it really doesn't matter. The genetic engineer discards off-types just like a conventional breeder would. Likewise, if I knew what four-letter word was being inserted in my text here, I could let the computer search it out and identify the random placement. I could discard the variations of the text that might alter the meaning on the resident sentence and keep that one version where the new inserted word makes the least contextual difference. One inserted word or gene is not a transforming force. Nonetheless, biotechnologies are currently being developed to precisely insert genes into exact chromosomal locations, and that feat can be expected to improve the predictability of insertion.

It takes multiple genetic changes to make most phenotypic changes. When we dig up and observe carrots, we notice that the roots of wild carrot and domesticated carrot look quite different and are the product of selected genetic variation (figure 2.4). However, wild and domesticated carrots still have enough genetic similarity to interbreed with one another. The genetic changes in domesticated carrot, while dramatic, were still not sufficient to convert it into a different species from the wild progenitor. The same can be said for cabbage, kohlrabi, broccoli, and cauliflower—

all members of the species *Brassica oleracea*. Conversely, transgenic plants are identical to the parent plant except for one gene and one trait. The next chapter shows how this feat is technically accomplished.

Unintended Consequences

Many opponents of genetic modification claim that there could be an untold number of unintended consequences of releasing GM plants into nature. While I agree that nature is a very big place with many incompletely understood ecological interactions, I disagree that single gene introductions in GM plants are likely to cause any ecological disruptions. The "vastness-of-nature" argument is based on the notion that we cannot hope to understand the full effects of introducing a GM plant into the environment. The same can be said for the introduction of new conventionally produced crop varieties. Still, many people argue that we ought to slow the introduction of new technology until we can be sure no harm will ensue; this is also known as the precautionary principle. But is the precautionary principle scientifically valid in this instance? While the social and political aspects of the GM controversy are beyond the scope of this book, the precautionary principle will be addressed as far as science will allow. Politics and sociology are not within the realm of science.

The key to understanding the potential environmental effects of GM plants is integrating the sciences of molecular biology (biotechnology), genetics, agronomy, and ecology. Much of this book follows the available trail of data. Like police investigators, we will question witnesses and follow up scientific hypotheses, which will allow us to determine whether today's GM plants are guilty or innocent of environmental harm. We'll also examine the next wave of GM plants and interrogate them as well. Before we examine biosafety sciences, though, let's see exactly how genetic engineering is performed and consider the changes in the plant that are wrought by stably inserting single genes into the plant genome.

3

Plant Biotechnology

The Magic of Making GM Plants

Central Dogma: One Gene, One Protein

The scientific principles behind biotechnology are anchored in what is known as the central dogma of biology and genetics. DNA, the double-stranded stuff that genes are made from, contains genetic information that is in turn transcribed into RNA, a single-stranded polymer similar to DNA. RNA molecules are then translated to produce specific proteins. This process occurs in each cell of a plant or other organism continually. Thus the flow of information is DNA to RNA to protein. Our knowledge of this process began with Watson and Crick's elucidation of the structure of DNA in the 1950s and has progressed until the present; by the 1980s, our understanding of basic molecular and cellular processes had created new biotechnology opportunities. The central dogma, DNA to RNA to protein, allowed scientists to predict the results of genetic engineering. If gene *X* was inserted into a plant, then the engineered plant would produce protein X. Genetic engineering bolstered the "one gene, one protein" hypothesis. The new biotechnologists believed that it would be possible to introduce a single gene of any origin into a plant and cause it to produce a single new protein. It was believed that introducing new proteins into plants would produce a better crop plant, helping farmers through increased efficiency and profitability.

Several analogies about DNA and genetics help people better understand their functions. The example I favor likens the bases in DNA, adenine, cytosine, thymine, and guanine (ACTG), to letters of the alphabet.

In ordinary language, letters form words, which have meaning to the reader. Genetic language is encoded by DNA and decoded by the reader—the cell. The bases ATG, for example, spell a word calling for the amino acid methionine. Each three-letter combination encoded by a DNA sequence is called a codon. Each codon spells out an amino acid "word," such as the ATG–methionine example. Amino acids are building blocks for proteins, and the amino acid sequence is crucial for protein function. One might define a gene as a genetic paragraph, which is composed of codons. The same is true for proteins.

Proteins have various functions within cells and organisms. Some proteins are structural, whereas many are enzymes that participate in biochemical reactions. Proteins perform various other vital cellular functions as well, and taken together, they produce discrete organs and tissues with unique properties. If we want to introduce a single protein-based trait in plants, the perfect tool is genetic engineering. Genetic engineering inserts a novel gene into a plant genome (the collection of the plant's genes) within a single cell. The new gene is called a transgene, and the new protein is called a recombinant protein, transgenic protein, or overproduced protein. The genetically engineered cell is called transgenic. Tissue culture techniques can be used to regenerate an entire, fully fertile and functional transgenic plant from a single cell. When the plant reproduces, it will yield transgenic seeds. "Transgenic," "genetically engineered," and "genetically modified" are all synonymous and used interchangeably. Because humans have spent millennia performing genetic modification of one sort of the other, I prefer "transgenic" as the more descriptive term when referring to genetic engineering.

Genetic Modification History: Tissue Culture

Like many other technologies, the biotechnology of genetic engineering has grown by leaps and bounds since the 1980s. Plant genetic engineering was only beginning to be understood and practiced in the early to mid-1980s. In the early 1980s, researchers had just begun devising methods to identify, isolate, and reproduce (clone) genes from plants and other organisms such as bacteria and even animals. Genes could be maintained in a test tube outside of their host organism. Between 1983 and 1985, the biotechnology was sufficiently developed to allow the transfer of foreign genes into the genome of plants for the first time. In the beginning, only a few plants were amenable to this technology. One of these was tobacco, and it has proven to be an exceptional transgenic model.

Tissue culture is the technique used to manipulate sterile plant cells and organs *in vitro* (literally, "under glass"). Until recently, this technique has been absolutely necessary for genetic engineering (methods to engineer intact flowers have now been shown effective for some plant species). Lidded, sterile plastic dishes containing a gelatinlike substance with plant food are used to grow plants and pieces of plants under sterile conditions. Sterility is important because bacterial or fungal contaminants will kill the plant tissue and ruin an experiment. The earliest plant tissue culture experiments, performed more than 50 years ago, demonstrated that tobacco cells and tissues could be successfully cultured. It was found that, by altering the plant hormone levels in the sterile media, shoots and roots could also be produced from undifferentiated cells in a petri dish. These new plants, originally springing from single plant cells, were genetically identical to (clones of) the originals. Tobacco seemed to be an ideal plant for tissue culture; it was hardy, easily cultured, and there were no obvious mutations because the new cloned plants looked, grew, and reproduced perfectly normally. Most important, genetic engineering was easily achieved with tobacco.

Tissue culture biotechnology has proven to be extremely valuable, even apart from genetic engineering. One of its uses has been the propagation of valuable plants, such as orchids, because it allows the recovery of multiple clones very quickly. Tissue culture also enables the study of basic cellular functions of plants *in vitro*.

Researchers have used tissue culture in mutational breeding to improve crop plants. In this process, cells are mutated using radiation or chemicals, and tissue culture regenerates selected plants with novel traits. For example, aluminum tolerance in plants would be a beneficial trait because that metal is toxic to the roots of most crop plants. Aluminum is among the most abundant metals in the earth's crust, and it is a big problem for many farmers; when roots are killed, plants are highly vulnerable to drought. Using tissue culture, a researcher can place mutated plant cells in media containing aluminum. Those cells that survive must have a mutagenized gene or genes that detoxify, sequester, or exclude aluminum from root cells. A plant that is regenerated from the tolerant cells stands a strong chance of being aluminum tolerant as well (figure 3.1). This ability to recover whole plants from single cells, called totipotency, is one of the basic necessities of the genetic engineering of plants.

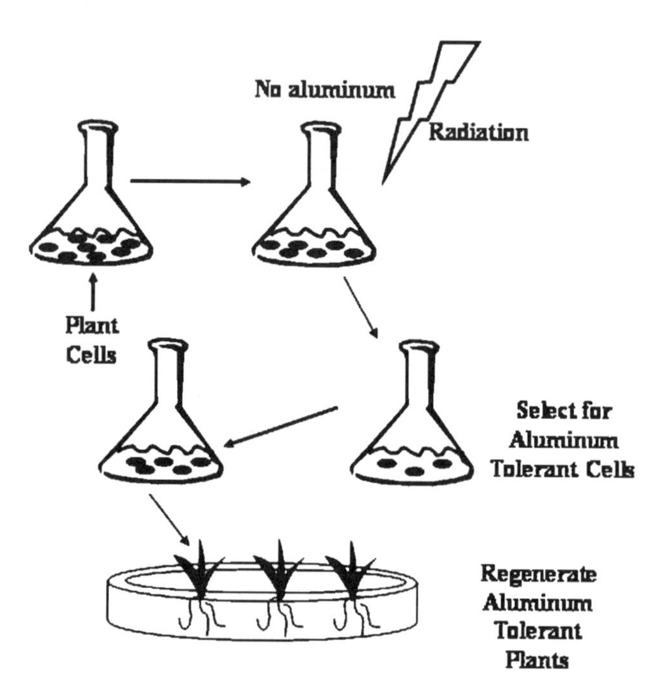

Figure 3.1. Flow diagram of tissue culture for aluminum tolerance. Cells are mutagenized by fast neutrons or by chemicals that produce random changes in DNA of the plant, and aluminum-tolerant cells are selected and plants regenerated. (Figure by Sandy Kitts.)

But before we get to genetic engineering, let's briefly discuss mutational breeding. Mutational breeding is routinely accomplished using methods that randomly change DNA. Many genes are altered, and the resulting plants often have unduly changed properties. Mutagenesis is a crude tool for modifying the plant genome that is completely unregulated by any governmental officials; genetic engineering is orders of magnitude more precise but is highly regulated by all governments. The dichotomy between precision and regulation of the two technologies is mysterious.

More History: Plant Transformation

More than 30 years ago, several researchers began studying how crown gall disease is transmitted to plants. Crown gall is a type of plant tumor that appears on many broadleaf (dicotyledonous) plants. It was found that the disease-causing agent, *Agrobacterium tumefaciens*, a soilborne bacterium, was capable of transferring some of its genes into the host plant genome at wound sites by natural processes. *Agrobacterium* makes plants produce sugars and opines (amino acid derivatives) needed for the bacterium's survival. Unique in nature, *Agrobacterium* naturally engineers the plants to pirate the plant's metabolic cellular machinery for the sole purpose of nourishing the *Agrobacterium*. We can think of *Agrobacterium* as a very smart (freeloading) micromachine that makes its living by invading plant cells, then reprogramming the plant's software (DNA) to make the plant feed it. Through many years of research we've acquired the genetic blueprints of the micromachine, as well as its wiring diagrams. We have intensely studied its invasive actions, and we've learned to reconfigure *Agrobacterium* into a controllable and predictable genetic engineering tool.

Agrobacterium, like all bacteria, possesses a number of plasmids, circular pieces of DNA that exist outside the bacterial chromosome. Plants, on the other hand, do not harbor plasmids. The most important plasmid for plant transformation is called the *Ti* (tumor-inducing) plasmid, found in *Agrobacterium*, which has several features that enable it to transfer its own genes into plant genomes. Genes to be transferred are located in a very specific region of the *Ti* plasmid called the T-DNA (transfer DNA). The T-DNA is bordered by DNA sequences that are mirror images of each other, and are called, appropriately, the right border and left border (figure 3.2). Copies of the DNA sequences between the borders are transferred into the plant genome, where they are integrated into a plant chromosome (figure 3.3). There is also a series of *vir* (virulence) genes that have been characterized and found to contribute to the *Agrobacterium*'s invasiveness, one property of which is the ability to transfer the T-DNA into the plant genome. Various *vir* gene-encoding proteins (i.e., VirA through VirG) work in concert to replicate the T-DNA outside the plant cell, produce single-stranded DNA, and then shuttle single-stranded T-DNA via a DNA–protein complex (called a T-complex) through the plant cell wall, cell membrane, and into the nucleus. Other Vir proteins integrate the T-DNA into the plant genome.

Originally, researchers cut and pasted the desired gene (the new transgene) into the T-DNA of the *Ti* plasmid. They quickly found that the huge *Ti* plasmid was very clumsy to work with. But it was discovered that the *Ti* plasmid could be divided into two smaller plasmids and still function normally. One of these, the helper plasmid, always resides in

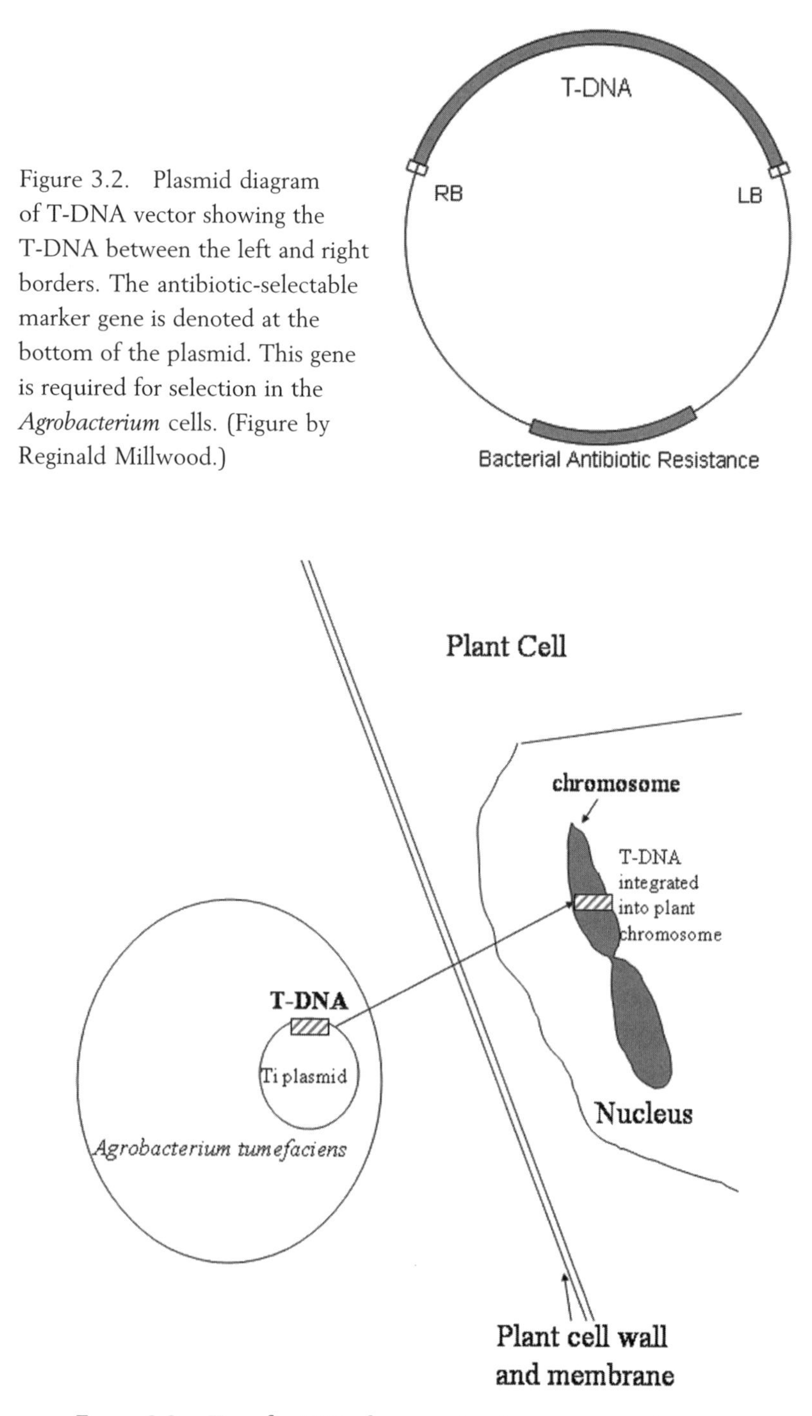

Figure 3.2. Plasmid diagram of T-DNA vector showing the T-DNA between the left and right borders. The antibiotic-selectable marker gene is denoted at the bottom of the plasmid. This gene is required for selection in the *Agrobacterium* cells. (Figure by Reginald Millwood.)

Figure 3.3. Transfer DNA from an *Agrobacterium tumefaciens* *Ti* plasmid being integrated into a plant chromosome.

the engineered *Agrobacterium* cells used for genetic engineering. It contains the *vir* genes. The other plasmid is called the binary plasmid; this is much smaller and easier to manipulate and contains the T-DNA. Today, researchers clone the transgene of interest into T-DNA of the binary plasmid leading to the subsequent plant genetic engineering using the *Agrobacterium*.

Gene Construction

How might a useful transgenic construct be built on a plasmid? The choice of transgene is myriad. The transgene might originate from any number of donor organisms and confer a wide range of potentially useful traits to its new transgenic host plant. Some of the more popular genes to put in plants have been isolated from *Bacillus thuringiensis* (Bt), a soilborne bacterium producing toxins that kill insects. Bt endotoxin proteins kill insects by making their guts leak. We will examine the mode of action and the importance of Bt endotoxin genes in subsequent chapters.

Once we have a transgene candidate in hand (in tube), we must clone it into an *Agrobacterium* binary plasmid. Before the Bt gene is subcloned into the T-DNA, though, molecular manipulations must be performed to ensure that the Bt gene will produce Bt protein. The gene must be flanked by the appropriate regulatory DNA regions that will eventually cause the plant to express the transgene (to make the specific protein). The two most important genetic regulatory regions are the promoter and terminator. The regulatory regions and the transgene are all made of DNA, although genes contain codons while regulatory regions do not. The promoter, the more crucial of the two, is located upstream, or at the front of the transgene. The promoter acts as a type of gene switch that turns gene expression on or off. One type of promoter, called a constitutive promoter, controls a gene to be switched on in all the tissues all the time, ensuring that the recombinant protein is continually produced everywhere in the plant. The constitutive promoter used most often in transgenic plants is the cauliflower mosaic virus 35S promoter (CaMV 35S, or simply 35S). Strong constitutive promoters are especially convenient to use because the transgene will be expressed in all parts of the plant during its complete life cycle. Most normal plant genes (and animal genes), however, are not controlled by constitutive promoters; this fact is illustrated by contemplating the various proteins that do specialized jobs in specific human cells. For instance, red blood cells primarily produce the protein hemoglobin, whose job is to carry oxygen. Hair follicles don't produce hemoglobin, but rather produce keratin, the hair protein. There are also promoters that have tissue-specific or induced expression patterns. The discovery and molecular characterization of promoters is a high research priority.

The terminator is the regulatory region located at the back or downstream of the transgene. It guarantees that the messenger RNA, which is the intermediate step between genes and proteins, is processed properly. Without the promoter and terminator, recombinant protein production cannot occur in the transgenic plant.

After we have a gene construct with an appropriate promoter, transgene, and terminator, we are getting closer to producing transgenic plants. But we need more. We will need to add an antibiotic resistance gene or a gene that codes for a protein that confers herbicide tolerance, and the appropriate promoter and terminator for these genes as well. Our binary plasmid map, a schematic of the construct with the two cassettes that will be transferred into plants, will look something like figure 3.4.

Plant Engineering Process

The binary plasmid shown above can be transferred into an *Agrobacterium* strain that is effective at infecting plants and is also disarmed—unable to cause disease on the plant. The genetic engineer wants the *Agrobacterium* to simply insert the transgene of interest and antibiotic resistance gene (all within the T-DNA) into plant cells. After that, the *Agrobacterium* is killed off.

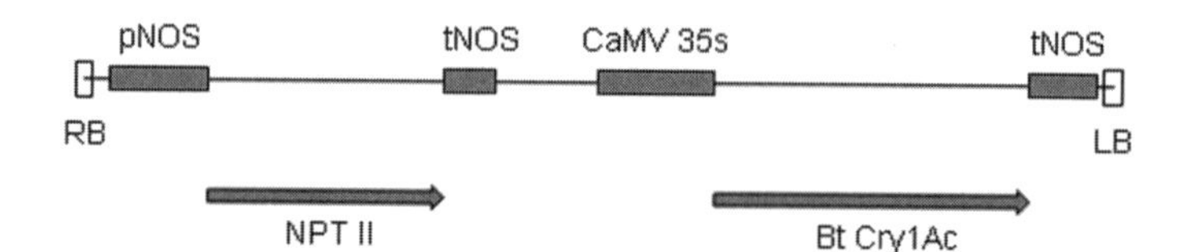

Figure 3.4. Plasmid map of the T-DNA region of an actual plant transformation plasmid. The Bt *cry1Ac* gene is under the control of the CaMV 35S promoter. The kanamycin-selectable marker gene, *nptII*, is under the control of the nopaline synthase (nos) promoter. Both genes have nos terminators. (Figure by Reginald Millwood.)

The process of gene transfer or plant transformation will be illustrated with tobacco. Plant molecular biologists use tobacco as a genetic engineering plant tool and do not consider it a crop (the U.S. government doesn't fund researchers to improve tobacco as a crop). First we need an explant source, a source of cells that we can transform. In this case it will be dime-sized leaf discs cut from intact tobacco leaves. The leaves come from plants grown from seeds entirely in a sterile medium that contains every essential vitamin and nutrient. The medium has the consistency of firm gelatin, and it contains sugar, so that photosynthesis is not needed for survival. Leaf discs (the explant tissue) are cut from the sterile donor plant, then soaked in an *Agrobacterium* solution to allow the tobacco cells to be transformed. *Agrobacterium* cells cling to the leaf discs and go about their business of genetic engineering.

The leaf discs are then placed in petri plates with more medium (we'll call it medium A), but this time the medium contains auxin, a plant growth regulator. Auxins, small molecules with indole rings (containing six- and five-membered conjoined rings), exist naturally in plants, and their effects have been thoroughly studied in nature, as well as under tissue culture conditions for all kinds of plants. The auxin will cause the wounded cells at the edge of each leaf disc to grow undifferentiated tissue called callus. In 1 or 2 days the *Agrobacterium* cells will be actively growing on the leaf discs, growing colonies. If the *Agrobacterium* is too virulent and active, it will simply kill the leaf tissue; its growth needs to be terminated before it overgrows and destroys the plant tissue.

During the first few days of culture, the transgene of interest and the antibiotic resistance gene will have been transferred into the plant cells and genome. The agrobacteria's work, from our point of view, will then be over. The antibiotic (antibiotic 1) kills bacterial cells but not plant cells. Applicable antibiotics are carbinocillin, mefoxin, or timentin, which are often used to treat bacterial infections in people and farm animals. The incorporation of antibiotic 1 into the medium will ensure that all plant cells but no bacterial cells have a chance of survival. Ultimately, however, we don't want all of the plant cells to survive, just those that have been transformed. To select the plant cells with the transgene and the antibiotic resistance gene, we use a second antibiotic (antibiotic 2) that will kill all of the plant cells not expressing the antibiotic resistance gene (and therefore not expressing the transgene). Very often the plant biotechnologist will use the neomycin phosphotransferase resistance gene (*nptII*), which confers resistance to the antibiotic kanamycin, and then use kanamycin as antibiotic 2. Tissue culture medium A containing these two antibiotics will serve three purposes simultaneously: (1) undifferentiated callus tissue will grow from leaf tissue, (2) transformed plant cells will be selected, but (3) *Agrobacterium* cells will be killed off.

After 2–3 weeks, healthy plant tissue will be transferred to new media, medium B. Medium B will also contain the two antibiotics, but it will contain another plant growth regulator, cytokinin, as well. Decreased levels of auxin will be used in this medium. Cytokinins stimulate shoot production, and in a few weeks transgenic shoots will appear.

These shoots can be excised and transferred to a third medium, medium C, that will either have no growth regulators or a low concentration of auxins. Medium C will cause the transgenic shoots to produce roots. In a few more weeks, we will transfer the new little transgenic plantlets to soil and let them grow up in pots to adulthood where they'll flower and produce seeds through self-pollination. The transgene and the antibiotic resistance gene will, from now on, act exactly like plant genes. The transgenes will be passed through pollen and egg to seed onto progeny plants that will inherit the transgenes just as if they were plant genes to begin with.

The transgene is incorporated into a distinctive chromosomal location, and a unique transgenic plant is recovered. This independent transformant is called an "event." Multiple transgenic events are recovered in any particular experiment, and experience has shown that some events will express the transgenes to a high degree and some less. Some events will have multiple transgene copies in the genome, and some will have only one copy. Plant biotechnologists generally prefer to use transgenic events that have a single copy of the transgene because of the simple Mendelian mode of inheritance. Events that highly express the transgene are also typically desirable. Above all, we want to grow and use transgenic plants that appear normal and produce seeds to the same degree as the non-GM parent plant. Odd mutant-looking plants will be culled but may someday have some novel, curiosity value (figure 3.5). Sometimes, however, the transgene will be integrated within an already resident plant gene. If this occurs, then the interrupted plant gene will not be expressed, and this can have profound effects, usually visible during the initial characterization period of a GM plant. These off-types are discarded. Certainly, if the transgenic plant event does not produce as many seeds as the nontransgenic parent, it will also not be useful in agriculture. In this way, plant transformation is much like conventional plant breeding; only advanced types are used.

Figure 3.5. Photos of off-type canola plants (a) and normal GM and non-GM canola (b). The off-type leaves are bleached white with red tinges and are somewhat crinkled.

Before plants can be commercialized and grown as nonregulated agricultural crops, strict government regulatory requirements are imposed. Because obtaining normal-looking, normal-yielding plants that will highly express the transgene is so vital, biotechnologists will often produce dozens to hundreds of unique transgenic events to find the one perfect GM plant. It is important to note that the transgenic event is also the unit that gets deregulated. So although there have been more than 1000 Roundup Ready soybean varieties currently in production since 1996, there was, until very recently, only one deregulated event. This event was produced in a Monsanto lab in the 1980s. The varieties containing the transgene came from this single event that has been subsequently used in conventional breeding to move the transgene to other genetic backgrounds—elite soybean varieties and breeding lines.

When *Agrobacterium*-mediated transformation works, it works like a charm. However, not all plants are as easy to transform as tobacco. In fact, all important crop plants are more difficult to engineer. *Agrobacterium* does not tend to infect grain crops such as wheat, rice, and corn, even though *Agrobacterium* strains and improved methods have been recently developed that are effective at engineering these plants. Fortunately, we have another tool and process (biolistics) as an alternative to *Agrobacterium* to insert genes that will transform plants. That tool is called the gene gun. A Cornell professor, John Sanford, and colleagues invented the gene gun in the late 1980s.[1] Using compressed helium, the gun blasts tiny pellets coated with DNA into plant cells (figures 3.6 and 3.7). Biolistics, also known as microprojectile bombardment, is commonly used instead of *Agrobacterium* to transform plants. It has the added advantage of not requiring the special binary plasmid system or T-DNA. Many of the tissue culture steps described earlier could be similar to those regenerating gene-gun–produced transgenic plants, although antibiotic 1 would never be needed, as no bacteria are involved in the process. All commercially grown GM crop plants were engineered using either biolistics or *Agrobacterium*.

Research facilities that perform plant transformation experiments must meet certain governmental biosafety requirements. GM plants may be grown outside the lab only under regulated conditions until deemed safe by the U.S. Department of Agriculture (USDA) Animal and Plant Health Inspection Service (APHIS). APHIS must issue a permit or agree to a notification before GM plants may be grown outside. Other countries have similar regulatory agencies. It is only after GM plants are grown outside under real field conditions that researchers can accurately assess yield and transgene expression patterns (i.e., how the plants will grow in real life). Field testing is also necessary to study transgenic plant ecology.

Many countries have moratoria on growing transgenic plants in the field, even under stringent regulated conditions. Paradoxically, part of the rationale is that they claim that not enough information is available about the biosafety of transgenic plants in general or about a specific

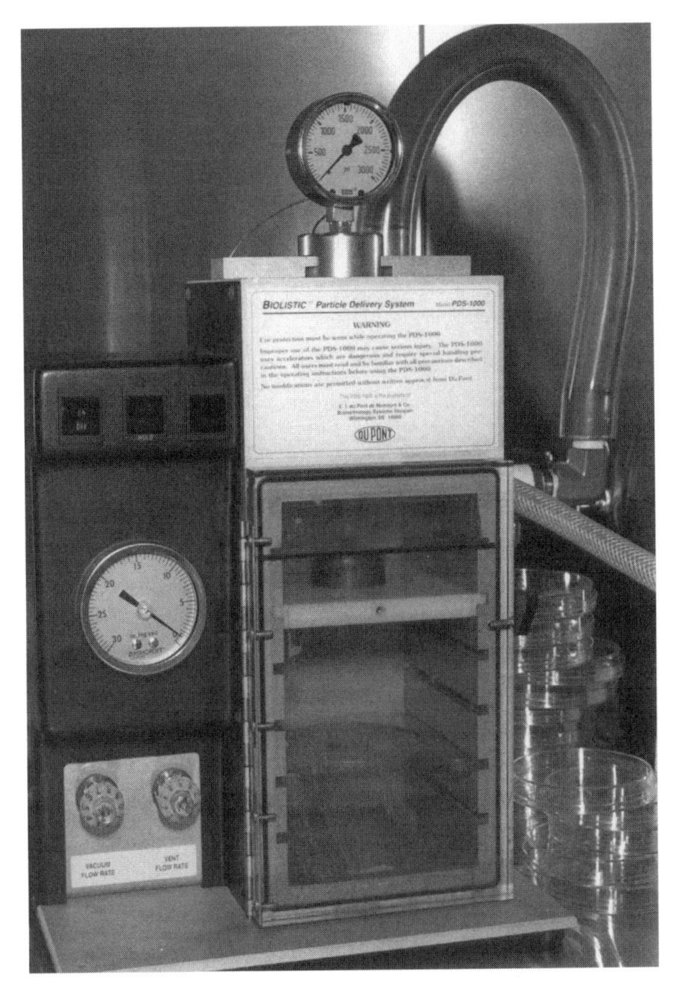

Figure 3.6. Helium-driven gene gun (biolistics) device.

transgenic event candidate (plant/transgene combination). This chicken/egg dilemma is unfortunate because the only way to assess the biosafety of a potentially useful transgenic plant with any accuracy is to grow it in the field, as we'll see in several examples in subsequent chapters. Growth chamber and greenhouse experiments work only to a very limited degree when one is attempting to characterize and predict how a GM plant event might perform in the field. A pilot can use a flight simulator any number of times, but eventually he will have to take to the air in a real airplane.

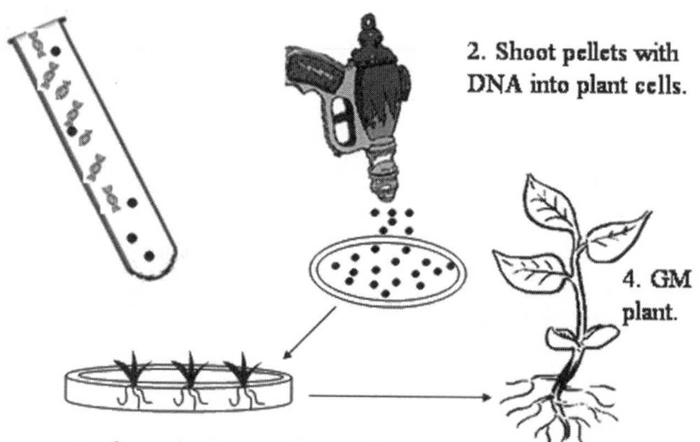

Figure 3.7. Schematic of gene gun-mediated plant transformation. (Figure by Sandy Kitts.)

Not Magic after All

College students who have taken my plant physiology course routinely and easily produced transgenic tobacco plants as part of a class lab experience. I have cooperated with several other college professors who have undergraduate students do the same thing as part of course requirements. It would be a relatively simple chore to develop plant transformation kits for high school students. Plant genetic engineering is a well-characterized process and one that is not difficult to execute. The making of transgenic plants is not magic, after all; their production simply uses the facts and tools of molecular and cellular biology to put novel genes into plants. Nature has given us an ample toolbox. The resulting transgenic plants appear as ordinary as their precursors and are typically indiscernible from their non-transgenic counterparts, and the transgenic trait is rarely readily visible unless you know when and where to look. If a GM plant has a Bt gene for insect resistance and is grown in a mixed population with non-transgenic plants, we could only differentiate which plant is transgenic if potentially defoliating target insects are present. The Bt plant would be obvious in contrast to defoliated parts of the non-transgenic plant. If there were no insects present, all the plants would look alike.

We have seen that there is a difference between crops and weeds and that genetic engineering has simultaneous aspects of biochemical precision (we can predict the trait or phenotype) and genetic randomness (we don't know where the transgene will land). We've also touched on the fact that ecological interactions are less predictable than genetics or traits. This brings us to questions that, because of my research, I get asked all the time. Will GM technology beget superweeds that will harm the environment? Can creating a supercrop lead to the inadvertent creation of a superweed?

4

Gene Flow

It's a Weed, It's a Transgene, It's Superweed!

"There is no doubt that transgenes will flow from certain crops to their wild relatives. In certain cases, I'm sure they already have." The Greenpeace and Green Party lawyers glanced at one another with the tiniest of smiles at the words of my testimony. No doubt, they thought, their battle had just been won.

New Zealand Royal Commission on Genetic Modification

During 2000–2001 New Zealand convened a Royal Commission on Genetic Modification to comprehensively study the pros and cons of growing GM plants in that country. The Royal Commission was charged with determining the role transgenic crops would have in the country's agricultural future and determining if New Zealand should even allow GM plants under cultivation. I had been invited to give testimony to the Royal Commission as an expert witness in GM plant ecology. Under cross-examination from the two lawyers, I had just honestly given my opinion about the bottom line of gene flow: yes, it would happen.

While touring and speaking to various groups in New Zealand, the comparison between exotic genes (transgenes) and exotic invasive plants kept cropping up time and time again. Recall from chapter 2 that it takes more than one or two genes to make a plant an aggressive weed. But there seemed to be a number of people in my audiences in New Zealand who kept comparing GM plants to exotic and weedy introduced species, as if to predict corresponding ecological disaster if New Zealand were to grow GM crops. There was sizable worry that transgenes flowing from crops to weeds would create superweeds. There were at least three contributing factors that led to the comparison between exotic invasives and transgenic plants. First, New Zealand has a "clean green" marketing image. I passed a clothing store in a large New Zealand city that had a big sign in front advertising "Clean and Green Clothes from Nuclear-Free New Zealand." I'm still unaware of the direct connection between the absence of nuclear power plants and clothes manufacturing in New Zealand, but I can rest easy knowing my new sweater won't glow with radioisotopes. The clean, green marketing niche is indeed fruitful for commerce, especially export goods. Some New Zealanders view biotechnology as a threat to green marketing and branding.

The second factor revolves around critical problems with exotic species that have caused extensive economic and ecological damage to the country. The English colonizers made the mistake of trying too hard to make the island nation resemble home. Most of the notable pests are ground-dwelling mammals, all of which were introduced; there are no native terrestrial mammals in New Zealand other than bats. Invasive possums, feral cats, and weasels have decimated the native flightless bird populations. Kiwi birds, among others, are seriously threatened—a dire ecological disaster. People in New Zealand are very aware that these exotic species are bad news for native species.

The third factor is a political derivative of the first two. Certain environmental and political activists in the country have leveraged citizens' fear and knowledge about the destructive nature of invasive species. When coupled with the mystification of biotechnology, activists have successfully imprinted the connection between introduced species and introduced genes (figure 4.1). It was comforting to me as a scientist that the Royal Commission seemed to clearly understand the scientific basis between species and genes. It was also clear that each side was having its day in court and that all opinions were heard to derive a rational recommendation about growing GM crops in New Zealand.

After the yearlong fact-finding mission was over, the Royal Commission decision came down on the side of science. The commission cautiously recommended a move toward the development of using GM plants in agriculture. The GM plant moratorium has been lifted, and the government is examining its regulatory structure in light of the science presented during the Royal Commission. It is still unknown how the recommenda-

Figure 4.1. Activists protesting genetically modified plants. (Photo by Doug Powell.)

tions will be implemented. However, one of the concepts that was soundly laid to rest during the commission hearings was the myth of the superweed. Gene flow does not equal superweeds.

What Is a Superweed?

In the introductory chapter, I mentioned my personal journey from being the biosafety skeptic to holding a position of cautious acceptance. All the reasons for my change of heart are scientific. Since 1993, many GM field studies from around the globe have been published, and our knowledge has grown by leaps and bounds. Studies examining the biosafety of current crops have found few, if any, special risks in the offing. In most cases GM plants look and grow exactly like their non-GM counterparts, with the exception of anticipated effects rendered by specific traits; plants engineered with an insect-resistance gene, for instance, will kill insects. And the proof is in the eating: since 1996, 38 trillion GM plants have been grown in the United States with no corresponding ecological ill effects. The reasonably realistic worst-case scenarios scientists have mustered to envision have not been all that bad. One would be hard-pressed to find a reputable practicing scientist predicting an ecological disaster for a realistic deployment of a transgenic crop. Bioterrorism and stupidity don't count. For example, it is known that genes flow from sorghum to its terrible related weed, johnsongrass. It would be stupid to put an herbicide-tolerance gene into sorghum knowing that it would escape into johnsongrass, making the awful weed even more difficult to control. As far as I know, no scientist or company has even considered it.

Quite frequently I hear the term "superweed" tossed about. The word did not arise from the scientific literature, and scientists don't use the word to describe a natural or GM occurrence (except maybe as a bad joke). Some people equate transgenic weed with superweed. But would a weed that is transgenic necessarily be a superweed? There has long been a concern that creating a plant through biotechnology and having the transgenes escape from agriculture into nature is a scary thing, especially if the transgenes move from crop to weed. In certain circumstances such a jump might be feasible because there are wild relatives that are interfertile with crops that have been engineered. And as I predicted at the Royal Commission two years prior, the first case of a transgenic crop × weed hybrid arising unintentionally in the field has now been documented.[1] An herbicide-tolerance gene has been found in hybrids of field mustard, a relative of canola. As we'll see later, these weeds aren't so super—they can be easily controlled. I suspect that one reason I hear "superweed" so often is that my lab has been the first to move an insect-resistance transgene into a weed—also field mustard—on purpose. But we shall investigate if they are really super or as one writer has called them, simply, wimps.[2]

Worst-Case Scenario

When I was initially thinking about creating and then testing worst-case scenarios with transgenic plants, I let my mind wander through the following milestones:

1. Transgenic crop is produced and released.
2. Genes flow from transgenic crop to a wild, free-living relative (not a crop).
3. The nature of the transgene, the recipient plant, and environmental conditions combine to create a weedier weed.
4. The new plant has higher fitness, is more competitive, can occupy other niches, can displace members of its own species, and can even invade ecosystems: it is a superweed, a more invasive plant in nature.

Superweed Redefined

Some people might not agree with my superweed scenario. Some people would argue that the superweed could simply be an enhanced agricultural weed. Perhaps the newly discovered herbicide-tolerant field mustard could be defined as a superweed. So, before I describe work from my lab's superweed candidates, let's discuss herbicide-tolerant plants that that could have enhanced weediness in agricultural settings but not be invasive outside of agriculture. In this latter scenario the following steps might take place:

1. Transgenic crop for herbicide-tolerance is produced and released.
2. Genes flow from transgenic crop to crop or from crop to a related species that is not a crop.
3. The herbicide-tolerant crop or wild relative becomes more difficult to control in farmers' fields using the herbicide of choice.

The simplest definition of a weed is a plant out of place. So a sunflower could be a weed in a cornfield. And a canola plant could be a weed in a wheat field. These kinds of weeds are not often present in large numbers and do not have the weedy plant tendencies listed in chapter 2. Crop "volunteers" are generally not economically important weeds because they are relatively easy to control. Noncrop weeds are the worst weeds. And the very worst weeds are called noxious weeds. The USDA has a noxious weed list (http://www.aphis.usda.gov/ppq/permits/fnwsbycat-e.PDF), which is a roster of the most terrible weeds in America. There are special regulations about the movement of plant species on this list.

Some people have argued that transgenic crops for herbicide tolerance can become special weeds; that is, herbicide tolerance in a crop might make the crop a weed where it wasn't one before. A case in point is an occurrence in Alberta, Canada, in which a farmer discovered canola on his farm that was resistant to three herbicides. Some have cited this as an example of a superweed created by biotechnology. First of all, how did it occur? In 1997 a farmer planted canola in two fields. On one side of the road he planted Quest, a Roundup Ready canola. On the other side he planted canola tolerant to Liberty herbicide. He also planted a non-GM variety that is tolerant of Cyanamid's Pursuit and Odyssey herbicides. Three mistakes: all canola, all together, all in the same year—not recommended as good practices. But as I understand it, the farmer was performing an experiment to assess the possibility of transgene stacking. As any crop or weed scientist could have predicted, the experiment's positive outcome was no surprise. The next year the farmer discovered newly sprouted canola resistant to Roundup volunteering in fields where none had been planted, and indeed, even two different herbicides could not kill the plants. In 1999 triple resistance was confirmed, as quickly as it could have possibly happened.[3]

Some people say that this new accidental canola is a superweed because it cannot be controlled by the herbicides of choice, such as glyphosate (Roundup) and glufosinate (Liberty). But is the triple-resistant canola indeed a superweed? It is still canola, a crop that may have a few weedy characteristics, but that now has resistance to several herbicides. The farmer could have (and did) spray the fields with 2-4,D, a common herbicide, to completely eradicate the canola plants. The end effect is that this putative superweed no longer exists. It doesn't sound so super to me.

"Aha," I hear you say, what if those genes move from canola to a real weed, a plant that is not a crop, like the weed *Raphanus raphanistrum*, the wild radish. Or what if the weed was not quite as bad as wild radish—say, field mustard or wild turnip, *Brassica rapa*? Wouldn't that be a superweed? The answer is a definite "maybe." If a glyphosate-tolerant gene were transferred into a weed from canola, it would certainly be impossible for the farmer to control in his fields with glyphosate (figure 4.2). Transgenic weeds would be created through hybridization and introgression (introgression occurs when successive multiple hybridizations [backcrossing] with the wild relative and the crop genes [or transgenes] are stably inherited in the new plants, in this case, a weed). But the farmer could control it with glufosinate or some other herbicide, and the ersatz superweed would be finished.

Dr. Suzanne Warwick and colleagues in Canada discovered the glyphosate-tolerant hybrids between canola and field mustard.[4] In addition to the primary hybrids they have discovered plants, more field-mustardlike in appearance, that are likely backcrossed hybrids. Hundreds of these glyphosate-tolerant plants were found in a pumpkin patch in eastern Canada where unwanted canola volunteer plants had become established. The pumpkin farmer had exclusively used glyphosate as the herbicide to control weeds (he covered his pumpkins), thereby selecting herbicide-tolerant biotypes that happened to include primary crop × weed hybrids as well as possibly backcrossed hybrids. Since the Canadian re-

Figure 4.2. Canola field in full bloom. (Photo by Photos.com.)

searchers have been looking for herbicide-tolerant weeds for only a few years, it is impossible to extrapolate beyond the few fields found already. More research and field censuses are needed. But it is interesting that the large numbers of herbicide-tolerant field mustard hybrids were found on a farm where glyphosate was extensively used for weed control.

To prevent the rapid emergence of herbicide tolerance and for multiple other reasons, farmers rotate their crops and alter weed control systems. If a farmer planted glyphosate-tolerant canola in the same field, or adjacent fields, year after year, then one might expect that glyphosate-tolerant weeds might appear that would become hard to control with glyphosate, as noted in the case study above. However, if the farmer plants a different crop or a crop with different herbicide-tolerance genes in subsequent years, he has created a moving target for tolerance and evolution. So it is not a good practice to plant glyphosate-tolerant soybean the year after growing a glyphosate-tolerant canola crop. During year 2, canola and wild relative volunteers that are glyphosate-tolerant would be selected for by glyphosate applications on the soybean. The whole purpose of planting glyphosate-tolerant crops to begin with is the enhanced ability to obtain efficient weed control. If the weed control system is misused and glyphosate-tolerant crops are overused, then the farmer is back where he started from: weed difficulties with no good way to control them. We will see an even more important risk arising from the practice of using GM herbicide-tolerant crops as a silver bullet for weed control.

In my opinion, the herbicide-tolerant transgenic superweed is a myth. It is hard to imagine a weed derived from biotechnology that possesses tolerance to every known herbicide and tillage practice. Herbicide tolerance as a trait can be fought with weapons farmers already possess. So let's look to traits other than herbicide tolerance when attempting to glimpse into the world of the superweed.

It is possible, I suppose, that an evil farmer or agri-terrorist could diabolically create a superweed containing transgenes that might make it more competitive in the wild. The superweed could then be unleashed. Or a scientist could make a superweed candidate in the laboratory and test it outside for scientific purposes. That was my goal.

Back to the Worst-Case Scenario

In 1994, I was considering worst-case scenarios to test. The requirements for an experimental system were (1) a crop amenable to transformation, (2) a crop that had weedy wild relatives subject to gene flow, (3) a transgene that coded for a trait conceivably beneficial to crop and weed alike, and (4) the entire package had to be relevant to agriculture. The transgenic trait was critical. Natural selection could favor it under the appropriate environmental conditions. For example, herbicide tolerance was not of interest in this situation because nature does not spray herbicides. The wild relative requirement was also important. Even in 1994 I seriously doubted that a single transgene would be sufficient to convert a crop into a noxious weed. My reasoning was that if the transgenes could move into wild relatives and the transgene conferred a trait that nature could select for, then perhaps the new plant might be ecologically invasive and disruptive in natural ecosystems—a good superweed candidate. Was it conceivable that genetic engineering could create the new kudzu? The goal was to make such a plant in the lab and test it in the field under regulatory scrutiny. It was reasonable to believe that in-depth experimentation with a worst-case scenario system would put all other less risky systems in perspective.

Which Crop?

The choice of crop to make transgenic was the crucial first choice. Most of the current transgenic crops in production in the United States, such as corn, soybean, cotton, and potato, don't really have wild relatives here. This is really good news in that these crops, the most popular ones, are safe with regards to gene flow and its consequences. In addition, these plants are not that easy to transform, despite the fact that there are plenty of transgenic varieties available. The crops with wild relatives in the United States that seemed to be the best candidates were rice, wheat, squash, canola, and sunflower. There are others, but these seemed to be the most reasonable to use as experimental models. Of these, only canola was easy to transform and had already been the subject of significant ecological biosafety research that would provide backdrop data; furthermore, canola seemed to be the most controversial transgenic crop candidate globally because it has wild relatives all over the place, not just in the United States. Canola, *Brassica napus*, also is a fairly new oilseed crop that is being grown in ever-increasing acreage in North America and elsewhere. It has a few weedy characteristics in comparison with more domesticated crops. While not a noxious weed, stands of canola can persist for a few years after cultivation in fields, waste places, and roadsides. However, canola does not really persist in unmanaged locations as sustained populations, so it was therefore a good model to test whether a particular transgene would enable the plant to maintain free-living populations. If so, it would prove that such a transgenic system is risky—that we could make a crop more invasive.

Which Transgene?

The decision of which transgene to insert was quite simple at the time, but there were any number of transgenic traits that could have been chosen as appropriate models. An appropriate transgene was simply one that could be expected to confer a fitness-enhancing trait to the host. Natural selection should "see" it. For example, a priori, traits that would enable a plant to tolerate drought, metals, diseases, or insects could be selectively advantageous, regardless of whether the host plant lived in an agricultural field or an unmanaged ecosystem. During the rest of the book, I'll refer to an imaginary gene that could give total resistance to everything: the *deathstar* gene. Of course, such a gene has not been discovered, but something like it was my goal. Essentially, the only traits I needed to stay away from were herbicide tolerance and altered oilseed traits. My transgene decision was made by convenience. I had been hired less than a year earlier to engineer soybean with a Bt *cry1Ac* gene at the University of Georgia. The *cry1Ac* gene product confers resistance to certain defoliating lepidopteran larvae (injurious worms), and the same gene had been thoroughly tested in other crops. There were certain ubiquitous canola pests that would be controlled by Bt, so it seemed like a good choice. The only real problem was that no canola transgenic with this Bt gene existed at the time, and I had to make the transgenic plants myself. Transgenic canola takes about a year to produce from tissue culture to progeny plants. I put together a multidisciplinary team to work on the project: Paul Raymer, a canola breeder and agronomist, and John All, an entomologist, both at the University of Georgia. Wayne Parrott, who had hired me as a postdoctoral fellow, was generous enough to allow me to use his lab to begin the work. We submitted a grant proposal to the USDA Biotechnology Risk Assessment Grants Program, and our Bt canola project was funded. We were off and running.

Making Canola a Weed

The initial experiments showed that the transgene was expressed in canola and killed target insects (figure 4.3). Insects would die a few hours after they fed on Bt plants. By their nature, Bt genes are not deathstar genes, but they have very specific modes of action and kill only certain types of insects. For example, Bt Cry1Ac toxin only kills certain lepidopteran larvae when ingested. The larvae of related species may not be affected whatsoever. And larvae of different insect orders (e.g., flies or beetles) would also not be harmed. However, Bt toxins are very toxic to target insects and have the capability to completely protect a plant from certain potential defoliators. In addition, there are other Bt toxins that control different insects compared with Bt Cry1Ac. For example, Cry3 toxins kill certain coleopteran (beetle) larvae. Because canola has only a few important insect pests and Bt Cry1Ac would kill many of these, the transgenic plants could conceivably gain a selective advantage against their nontransgenic counterparts. This fact did make Bt *cry1Ac* the closest thing to a *deathstar* gene we could find. It was a good fit to test the hypothesis.

When we put the Bt canola plants in the field in mixed populations with non-GM canola with potential defoliating insects present, like the cosmopolitan diamondback moth (*Plutella xylostella*) larvae, it was found that the Bt canola plants indeed produced more seeds and had a higher fitness compared with nontransgenic canola.[5] This, too, was not a big surprise. So, if free-living Bt canola populations could exist, even for a

Figure 4.3. Photo of non-GM canola (right) and Bt-transgenic canola (left) under heavy insect herbivore pressure. Note caterpillar inching up the remnant of the mostly eaten leaf.

few years, let's say on roadsides, potential defoliators could select plants containing the Bt gene. These Bt plants (really the Bt gene in the plants) would be selected for and persist in mixed populations of canola. The Bt gene would then be eventually fixed in populations; all plants in that population would have the Bt gene. It is important to note that we found that the Bt canola plants were not more invasive than the established old-field vegetation. Bt canola simply could not outcompete annual and perennial grasses, such as crabgrass.[5]

Population Genetics: A Quick Primer

How do genes get in and out of populations? What are the forces that control gene frequency? We will examine an animal example, but the same rules apply to plants as well. In some parts of the United States and Canada, there are black squirrels, but in most parts of these countries, the squirrels are gray. Let's say there is a single gene that controls squirrel fur color. There can be two variants, or alleles, for that gene: gray and black. And the gene exists on two homologous chromosomes: two chromosomes that have the same genes on them. It is typical for allelic forms to exist. You probably remember some of this information from the "Mendel's peas" lesson we all had in school. If Gregor Mendel crossed a short pea ("short" is the phenotype; the trait we observe; *ss* is the genotype) with a tall pea (*SS*), then all the resultant pea plants would be (*Ss*)—also tall. In this case there are two alleles for the height gene. If he then crossed the *Ss* peas with each other (parents are still all tall), then 75% of the pea children would be tall (*Ss* or *SS*; 2:1) and 25% would be short (*ss*). Homozygosity is the state of like alleles (e.g., *SS* or *ss*), and heterozygosity is illustrated by the *Ss* state. If squirrel color is transmitted this same way, then we might expect to see more genetic variation (and fur color variation) in individual populations of squirrels than we do. So why don't we see populations of squirrels where there is a mix of black and gray types? Why do we see homogeneous populations of either black or gray? To understand the lack of genetic variation in populations of organisms, we have to look at the rules of population genetics. If the squirrels had related species (they don't) to which they could transmit the fur color gene, these would also obey the laws of population genetics. Ditto for multiple genes, and so on. While our particular example is simple, the rules are the same for more complex cases.

Heterozygosity

If most individuals in a squirrel population have gray fur (GG), and there are just a few individuals with black fur (*gg*), then the allele frequency for gray would be nearly 100%. If one particular squirrel has at least one

G allele, then it will have gray fur (GG or *Gg*), and, therefore, we see that the G allele is completely dominant. Independently, scientists Hardy and Weinberg described mathematically how to calculate allele frequencies in a population with random mating. The following equation is known as the Hardy-Weinberg equilibrium:

$$p^2 + 2pq + q^2 = 1$$

In our case, let p be G and q be *g*.

Let's say that only 1 out of 100 squirrels is black (*gg*), $q^2 = 0.01$. That means $q = 0.1$ and then it follows that $p = 0.9$. And by solving for the Hardy-Weinberg equation, we'll know that there are 81% of the individual squirrels with the GG genotype (gray), 18% with the *Gg* genotype (gray), and we have known all along that 1% of the population is black (*gg*). So if we start with a rare allele in a population, such as putting in a transgene into a weed population, what forces will act on the allele (the transgene) to make it more frequent or cause it to decrease or even disappear? As we'll see, our example is overly simplified. We are pretending genes are independent islands within the genome. Genes are actually interconnected on strands of DNA called chromosomes. Aside from chromosome effects, there are four forces that will act on alleles: mutation (or transgene introduction, in our case), migration, selection, and drift.

Mutation

How do new genes and alleles arise in a population? Mutation is considered to be a first source of genetic variation. There might be a basepair change or arrangement in the DNA that allows a new gene to arise; in our case, new genes are introduced into populations using GM processes. No homologous (corresponding and similar) gene was present before the introduction of our GM trait. As in the pea and squirrel examples, the transgene is dominant (+), such as would be the case for our Bt gene. If it is present in one copy, termed heterozygous or hemizygous (+–); the term hemizygous is used where no homologue exists, such as in transgenic plants, but the two terms are synonymous), the transgene will be expressed in the phenotype. A homozygous state for the transgene (++) will also be expressed, but not a homozygous null (– –). In this latter case, the trait would be absent. Once the transgene is present (as a novel allele), will it persist in the population and move into new populations? Migration is the most important factor in transgene movement.

Migration

Genetic migration from one population to another population commonly occurs in a couple different ways. In one way, the seed can be dispersed by animals or by wind, or simply carried by gravity and running water into new populations. Populations might even be founded by transgenic plants, giving a high initial allele frequency. For example, transgenic seeds could fall off a grain truck and establish new roadside populations. A second avenue of movement could result from the transgene traveling via pollen from one population to another and being introduced via pollination. The exchange of genes (alleles) between populations is usually a function of distance and types of mating. Plants can mate by self-pollination, cross-pollination, or vegetative reproduction. It is important to note that it does not take very much migration to maintain an allele in a population. Migration is also called gene flow. Although gene flow is important in transgene movement, the most important force in maintaining an allele or transgene in a population is selection.

Selection

Selection, either natural or artificial, will increase allele or transgene frequency in a population as the fitness of the host organism increases. Fitness can simply be defined as the ability to pass genes on to the next generation. Accordingly, someone like the rock star Sting, who has six children (and who is, by the way, a proponent of population control), has a higher fitness than someone like me, who only has one child. It may be a putative rock star gene or the moneymaking gene in Sting that is selected for, or both. Selection, however, depends on the environment. You can think of the environment as doing the selection; it is like a sieve that allows some alleles to be selectively transmitted. For example, in an environment with lots of pest insects, a Bt gene might be selected for, and it will become more frequent in a population, but not more frequent in an environment with no insect pressure. In a society that values classical music, selection could choose against Sting's rock star gene, thereby decreasing his fitness. In an environment with no pest insects, a Bt transgene would be effectively neutral, neither being selected for nor against.

Drift

Genetic drift is a random process that is more pronounced in small populations. We can think of drift as the propensity for an allele to be lost or fixed in populations through random matings—sampling error. Let's think of allele frequencies of 0.5. Just by chance, a small population will deviate from 50% allele frequencies over time. As the allele frequencies start to move up or down, they will start rapidly sliding to either end. The allele will either be fixed in the population or lost, and drift, of course, would happen in the absence of selection. In reality all four forces play a part in population genetics. Transgenes are subject to the same forces as "natural" genes. Most studies that address the risks of GM plants focus on gene flow (migration) and fitness (selection). And once again, genes are assumed to be islands. Of course, we know that genes exist in linkage groups that are physically composed as chromosomes. The reason populations of squirrels tend to be either all gray or all black has to do with a violation of Hardy-Weinberg laws, which assume random mating. Gray squirrels prefer to mate with gray and black with black. Plants are not often so choosy—at least not on purpose.

Gene Flow to Wild Relatives

In our experiments, in addition to learning about gene flow and weediness, we also learned a great deal about insect behavior in a transgenic system containing Bt canola, how Bt canola competed against nontransgenic canola, and the fitness of Bt canola in the absence of insects compared to nontransgenic canola.[5–11] The next step was to investigate the movement of Bt genes from canola to wild relatives. We focused on the two prominent wild relatives of canola mentioned previously: *Raphanus raphanistrum*, the wild radish, and *Brassica rapa*, field mustard. They are generally found in canola fields, among other places. The initial experiments were hand-crosses, what we call "plant sex" to motivate the undergraduate labor who transfer pollen from a canola flower to that of another species. We observed that the cross with wild radish was difficult and the cross with field mustard was quite easy.[12] This, too, was not surprising, because similar results had been shown in the scientific literature. Lots of researchers had been able to hybridize and introgress genes from canola to field mustard, and only one group, a French group, had even moderate success in making backcrosses with wild radish.[13] But field mustard is a closer relative of canola than wild radish. The genetic relationships among *Brassicas* were worked out in the 1930s by the Korean scientist named U (figure 4.4).[14] In fact, canola contains an entire genome of field mustard. Canola arose millennia ago from a presumed cross between *Brassica rapa* (*AA* genome) and *Brassica oleracea* (CC genome). The *Brassica napus* genome is *AA*CC, and it is called an allotetraploid. Wild radish is in a different genus but shares some genomic homology to canola. While other researchers had moved genes from canola to these weeds, we were the first to move fitness-enhancing genes into hybrids and then to backcrossed hybrids. Were there any early surprises?

Transgenic Hybrids

We were not too surprised at the difficulty at making the wild radish cross, but we were surprised that the few hybrids we recovered using wild radish as a pollen recipient did not look like radish at all, but like canola (albeit

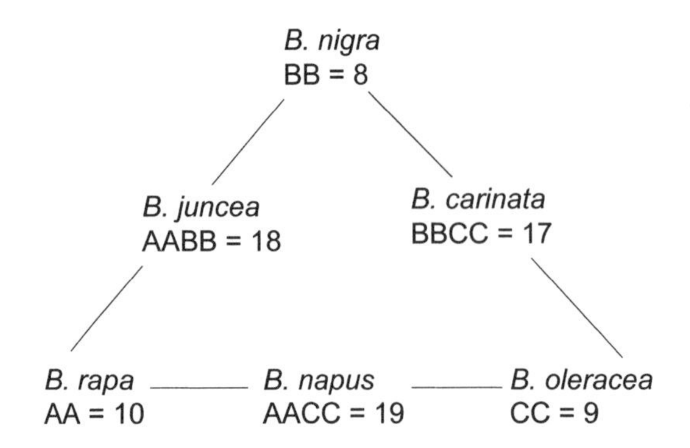

Figure 4.4. Triangle of U. The Korean scientist U hypothesized in the 1930s that certain *Brassicas* had specific genomic relationships with each other.[14] The numbers under species names are the haploid chromosome numbers. The letters indicate genome identity.

an odd-looking canola; figure 4.5). These plants, like canola, were self-pollinating and could not be backcrossed with wild radish and therefore represented a breeding dead end. The hybrids with field mustard, on the other hand, were plentiful and backcrosses were relatively easy.

The field mustard × Bt canola hybrids looked like hybrid plants, sharing characteristics of the two parents, and the backcrosses to field mustard looked increasingly like field mustard (figures 4.6 and 4.7).[12,15] After three crosses the backcrossed transgenic field mustard plants had the same number of chromosomes (20 chromosomes) as non-GM field mustard. At this point, we were confident that our Bt field mustard was the first insecticidal transgenic kind of wild plant—a transgenic weed (figure 4.7). During this time the research team had been expanded to include new researchers and collaborators. Matt Halfhill, a graduate student, was now performing most of the experiments (with his undergraduate helpers). Suzanne Warwick, the Canadian researcher whose team discovered glyphosate-tolerant field mustard in the pumpkin patch, was also contributing heavily to the project.

One of the first objectives was to determine if the backcrossed hybrids expressed the transgene, and they did.[15] Recall that transgenes are randomly integrated into the chromosome. In some chromosomal locations (events) the transgene is expressed to a higher degree compared with others. This is called the "position effect." Some transgenic events highly

Figure 4.5. From left to right: canola, canola × wild radish hybrid, and wild radish.

Figure 4.6. From left to right: leaves from canola, canola × field mustard (middle two leaves), and field mustard. (Photo by Matt Halfhill.)

Figure 4.7. From left to right: canola, field mustard × canola (F_1 hybrid), field mustard × F_1 hybrid (BC_1 hybrid), field mustard × BC_1 hybrid, and field mustard. (Photo by Matt Halfhill.)

expressed the transgene and effectively killed insects, whereas in others the Bt transgene was expressed to a lower degree or not at all and did not kill target insects. We crossed several different canola events with field mustard. In addition, we crossed the crop × weed hybrids with the weed to produce a series of backcrossed plants. All these hybrids and backcrossed hybrids had equivalent expression compared with the original Bt canola parent (event). That is not surprising, either, because when transgenes are transferred to another species through a sexual cross, flanking genes on the same chromosome are transferred, too; the position effect stays integral. The chromosome may be transferred intact, or chromosomal recombination could occur, so that a portion of the chromosome containing the transgene is transferred into the pollen recipient plant and subsequent hybrids and backcrossed hybrids.

Same as It Ever Was?

It is important to remember that genes are likely transferred between canola and field mustard continually in the field regardless of transgenic status. Genes have flowed between related species such as these for millennia. So what is the big difference between related nontransgenic plants having sex and a transgenic plant having sex with a nontransgenic plant? While the plant puritan might object on moral grounds, many researchers would argue there is no difference whatsoever. If there is one, and I believe there might be, it really has more to do with the particular transgenic trait. If a supertrait (e.g., Deathstar) were already in canola that might subsequently enhance the fitness of the wild plant, say insect or disease resistance, then there would be no fundamentally unique situation with transgenic plants versus nontransgenic plants. But there is a difference. Both wild and domesticated plants have genes that confer traits such as insect resistance, but the traits are usually polygenic (controlled by several genes) by nature and are relatively weak. A Bt gene, however, is a single gene that confers strong insect resistance. And that is why Bt genes are so useful in transgenic plants and important for agriculture. If the insect resistance inherent in crops were sufficient to prevent herbivory, then a transgene such as Bt would not be needed. In addition, many traits such as insect and disease resistance have already been conventionally bred into crop plants by using germplasm from other sources. One of these sources could be a landrace of the same species as the crop, and another popular source could be (surprise!) from wild plants in related species.

Wild plants tend to have traits that help the plants survive in nature where they don't have farmers intervening on their behalf. In fact, most of our knowledge about crossability between crops and wild relatives comes from years of crop breeders' research in attempting to make wide crosses for crop improvement. Not that all attempts at wide crosses are successful. It is important to note that failures are often not published. There is one other sizable difference between wild-to-crop gene flow (i.e., conventional breeding) and crop-to-wild gene flow (i.e., when trangenes escape): breeders impose "hard" selection on the hybrids to make sure the genes will be passed on to the new plants, while nature's selection is almost always softer.

Good for the Goose or Just the Gander?

Here's the rub: many people are worried that the transference of a Bt transgene or some mythical *deathstar* gene will make an ordinary weed a superweed. But if wild plants already have a significant degree of insect resistance genes through natural selection over time (one of the things that makes them good competitors), then how much would it help a weed to acquire a Bt gene? Would a Bt gene increase the fitness of the weed? It indeed might increase its insect resistance and fitness, but it will likely not help the weed as much as it improved the crop that did not have much insect resistance to begin with. Would it make the weed more invasive if the weed populations are not normally controlled by insects, but by other things? Could a Bt gene make a wild plant more competitive? These are all very good questions that will be addressed later in this chapter and chapter 10.

Outcrossing and Selfing

It is certain that transgenes can be transferred from canola to field mustard using hand crosses. But would the hybridization events occur in nature (e.g., without plant breeders helping the crosses occur)? For the crosses to potentially occur, the weed must grow in close proximity to the crop, and the crop and weed must flower at the same time. The plants could be growing together within the same field or the weed could exist at field margins or adjacent to fields. There are some data that suggest that pollen can travel miles, but how important is long-distance pollen dispersal? To answer questions about gene flow, we must know some basic pollination biology and population biology of the plants of interest.

Canola, like most crops, primarily self-pollinates. The pollen from one flower will fertilize an egg cell from the same flower. Sex within an individual is one of the big differences between plant and animal worlds. Crops have generally been selected for selfing so true-breeding plants can be easily obtained and multiplied. Weeds, on the other hand, tend to be outcrossers. They have self-incompatibility genes that ensure that a single plant does not have sex with itself, anti-incest genes that prevent inbreeding. What does this difference between crop and weed mating strategies mean for gene flow from crops to weeds? Let's imagine that a weed plant occurs very infrequently in a field of crops and that the crop and weed species were sexually compatible. The occasional weed plant would have a difficult time finding a mate of its same species (remember, it cannot self-pollinate). But there would be a high probability that crop pollen could fertilize the weed plant and lead to crop × weed progeny containing the transgene. Self-incompatibility of the weed would make the crop pollen the only game in town. The rare weed situation is common in agriculture. And weeds of similar biology to the crop (say, *Brassica* weeds in *Brassica* crops such as canola) are quite a frequent occurrence.

Field experiments have been carried out to simulate an infrequent weed situation in a sea of crops, where crop-to-weed ratios are around 600:1. Hybrids are created at a high frequency; sometimes as high as 30%. But what about the other 70% of the seeds? These seeds are harvested from the weed, which is supposedly self-incompatible (i.e., that 70% should not exist). As nature would have it, low levels of selfing go on in the weed after all, something that would not have been easily discovered if not for these kinds of experiments in which we have a very precise genetic marker (the transgene). When field mustard is rare in a field of canola, then gene transfer from crop to weed seems like a done deal.

But what if the crop is rare and the weed is prevalent? This might happen if some transgenic canola seeds fall off the turnip truck (a common occurrence as well) and land in a natural stand of field mustard. In this case canola pollen will have a much harder chance fertilizing a field mustard plant because the canola pollen will have to compete with loads of field mustard pollen for fertilization. The cross might be able to go the other way, however. Even though canola is self-compatible and will probably largely self-pollinate, if there is sufficient field mustard pollen, then the canola plant might produce some hybrid plants. Some hybrid plants have been found in field mustard populations in England that were growing adjacent to canola fields. But, in this situation weed × crop hybrids were very rare.[16] We have performed experiments to assess the frequency of backcrossing of the transgene into the weedy background, and that number is also much lower than predicted from lab experiments.

Sexual Dysfunction

After an initial hybrid plant is produced from an interspecific cross, it must be able to survive and reproduce. It will also probably be largely self-incompatible. If the cross is too wide and the species are too dissimilar, as we observed with the wild radish × canola cross, the hybrid might be largely sterile and a reproductive dead end. But if hybrids are normal-looking and normal-acting plants, like the field mustard × canola hybrids, then there would likely be no barrier for additional crosses with either canola or field mustard. So which will it prefer to mate with, and does it matter? It depends. If the crop is rare and hybrids are plentiful (which does not happen very often), then the hybrids would likely mate among themselves and not with the crop. This is the case because of the self-compatibility of the crop; most of the crop pollen is used on itself. The crop would have to be exceedingly lost in a sea of hybrids for the hybrid pollen to be able to compete with the crop pollen in fertilization events. But if the weeds were few compared with the number of hybrids, then transgenes in the hybrid would likely continue to flow to weeds. This would be a more prevalent situation than the prior one because the farmer tries to rid the field of weeds. If the hybrids are created outside of a farmer's field, then there is a strong likelihood they will persist—if they can compete with the other plants.

We have performed some field-level gene flow experiments to assess the rate of gene transfer when the weed populations vary. However, we believe there are no reasons for the Bt transgene not to persist in the environment, at least at low levels. After all, those plants that have a Bt gene can survive infestations of target insects. And recall that after three generations (one hybridization event and two backcrosses to the wild relative) we will have essentially a Bt transgenic plant that has weedlike chromosome numbers and many weedy traits. But persistence of a transgene in a natural environment depends on the host plants, whichever they may be, being competitive and fit—the consequences.

Consequences of Gene Flow: The Making of an Ersatz Superweed

The crux of the possible problem of transgene flow and persistence in the environment of transgenes in weedy hosts is not whether gene flow will occur or not. We know that hybridization will occur between canola and field mustard. Backcrosses to the weed will also occur, albeit at a significantly lower rate than initial hybridizations. Transgenes would flow from sugar beet to wild beet, sunflower to wild sunflower, sorghum to johnsongrass, rice to wild rice, squash to wild squash, and from oats to wild oats. There are numerous examples of crops that have weedy wild relatives.[17] The real question is what are the ecological consequences of the transgenes in weeds. Will the answer be a big "so what?" Or will a superweed be created? Or is the answer somewhere in between? The answer will likely be slightly different for each crop × weed combination that is examined, but there should be some generalizations that can be drawn from sound experiments in any crop–weed system.

The lessons we are learning with transgenes in canola and field mustard hybrids should serve as good models to predict how certain transgenes would act in other wild plants, if transgene introgression is possible or likely in those species. In lab experiments we have learned that certain insects, such as the canola pest (the diamondback moth), enjoy feasting on field mustard and wild radish equally as much as they do on canola plants. However, when we see these weeds in nature, there is little evidence to suggest that defoliating insects are important factors in limiting plant growth and seed production, and hence population growth and invasiveness. We know that at times populations of diamondback moth and other lepidopterans that eat these plants can be quite large. Little is known, however, about their effects on wild plants. Much of the defoliation in these wild plants is from snails and slugs, pests that we have no good transgenic solutions for. Additionally, slugs and snails are not significant economic problems in most crops, indicating that transgenic solutions will not be sought (figure 4.8).

Will GM Plants Decrease Genetic Diversity?

We still need to know more about the ecology, physiology, and genetics of weeds and invasive plants to make accurate predictions about the consequence of transgenes moving into weeds from GM crops. But do we

Figure 4.8. Snails and slugs damage leaves (top panel) of wild radish as well as siliques (bottom panel).

know enough to at least rule out the superweed scenario? I think the answer for most weeds is yes because single transgenes will likely not convert weeds into superweeds that could invade new ecological niches. Multiple transgenes are discussed in chapter 10, where risks are put in context.

One hot issue on the minds of biologists, agronomists, and other interested parties is the effect of gene flow from GM crops on genetic diversity. There are several reasons for concern. The first green revolution was built on improved crop varieties with higher yield—genes that increased effective crop fitness. The genes responsible for this improvement came from natural genetic diversity of crop varieties, landraces, and wild relatives. Presumably, these fitness-enhancing genes have flowed back into the wild relatives with no ill effects. Also, a high level of genetic diversity increases genetic and ecological buffering against perturbations, such as plant disease infections.

Let's consider two plant infectious diseases that changed the face of the world. The first is an agricultural disease, late blight of potato, which caused the Irish potato famine. The second, which occurred in a natural system, chestnut blight, decreased American chestnut from a dominant tree species in the Appalachian Mountains to an incidental understory plant (figure 4.9). In the first example, cultivated potato had a very narrow

Figure 4.9. Understory American chestnut saplings sprouting from ancient rootstocks in Virginia. The photo was taken in the 1980s, indicating that the dead chestnut log had been there at least 50 years. The cycle of sapling sprouting and death had likely been occurring since that time.

genetic base. Potatoes are reproduced clonally from the "eyes" on tubers. The potatoes that were decimated by potato blight consisted of one to a few different genotypes, and there was an absence of resistance genes in these potato clones. There are many additional examples of diseases that have severely impacted agricultural yield and socio-political situations in which added genetic diversity could have ameliorated the damage. But we are more interested here in outcomes in plant ecology. In this regard, the second example of an introduced fungal pathogen on chestnut is particularly interesting. While chestnut had plenty of genetic variation within and among populations, it simply lacked a resistance gene or genes against the exotic chestnut blight fungus. As we've seen, many people have compared transgenic plants to an invasive species, which is not a valid comparison, but there are related questions that are applicable.

Is it possible for a transgene to escape agriculture (yes), move to unmanaged populations or species (yes), persist (maybe), and then sweep through the population or species in such a way as to decrease the genetic diversity of the species? The last part of the question is the mother lode, even though we are not sure about the answer to the persistence question (which is discussed later). Even if our transgenic wild plant cannot be classified as a superweed, decreased genetic variation would not be desirable for either agriculture or nature. No one has performed the kinds of experiments with transgenic plants that would definitively answer this question for any transgene or plant. One significant reason that no such studies have been performed is that it would take a long period of time and lots of funding; two things that often don't go together. A typical federal grant is for $200,000–$300,000 for 3 years. Such a project would be well outside the scope of a typical grant. Nonetheless, what can we predict based on theory?

More Plant Sex

From our earlier primer on basic population genetics, recall that the four forces that drive population genetics are mutation, drift, migration, and selection. Here comes the midterm exam via example. Selection would be the force that could potentially decrease genetic diversity as a few fit genotypes are selected for. The other force that could drive genetic diversity lower would be genetic drift as alleles are randomly lost from small populations. Let's not be too concerned with drift because its effect is on small populations and we are assuming that a fitness-enhancing transgene will increase weed population size. What effect might selection have? It will primarily depend on the plant breeding habit. We will examine three potential breeding strategies that individual plant species might exhibit. Plants often mix and match these strategies. The simplest is clonal propagation, otherwise known as vegetative reproduction. Not many plants employ this strategy solely, but it is not uncommon in the plant world. Potatoes and cranberries are clonal. Cranberries are essentially wild plants that have been propagated by cuttings for hundreds of years. Clonal plants such as cranberries[18] generally have low genetic diversity because large individual clones might occupy large areas, and genes are not mixed from generation to generation via sex. Plants that are self-compatible and that primarily self-pollinate also have low genetic diversity. Soybeans, crop and wild versions, are an example of a plant that is largely self-pollinating; soybean breeders often have a tough time finding useful genetic diversity to breed into commercial soybean varieties because of the generally low genetic variation in the species. Self-incompatible plants (outcrossers) have the greatest amounts of genetic variation. They are continually exchanging genes with one another and recombining genes within and among populations.

Let's imagine that we introduce a fitness-enhancing transgene into each of these three mating types. Since we're familiar with our Bt gene and it is an appropriate choice for agricultural applications, we'll choose it as our example. A Bt gene conferring insect resistance would protect the host plants from injury caused by insect defoliation that they would experience in the absence of the transgene, and, as a result, the GM plants will have greater reproductive fitness when insect pests are present. In the first case, a clonal genotype with the transgene would become more prevalent in a population and perhaps outcompete all other clones over time. In this case, an already low genetic diversity would become even lower. The same outcome would be expected for self-pollinating plants. Unless the seeds can be widely dispersed, a local population might become more genetically homogeneous, but because genes are not flowing large distances, the effect will be localized in ecological time. In evolutionary time, both clonal and selfing species might expect to have decreased genetic variation as the strong fitness-enhancing gene leads to a selective sweep. But recall that most weedy and wild plants are self-incompatible and must mate with another plant or, more specifically, with a different self-incompatibility allele-containing plant. With clones and selfers, the genes are essentially inherited as a package, with the transgene theoretically driving selection and carrying all the other linked genes in tow. For self-incompatible plants, the transgene and surrounding genes on a chromosome will be the selective package, not the entire genome. In this case, the transgene should not lead to decreased genetic diversity in the genome of the wild plant. Rather, the insecticidal transgene could become fixed in the population and spread to other populations. Other genes, except the genes that are closely linked to the transgene, would be largely unaffected. This is because outcrossers shuffle the genetic deck at a higher rate than selfers. Therefore, we should not worry much about decreased genetic diversity because most of the wild and weedy plants are outcrossers. But this brings us to an interesting proposal to control the fixation of a fitness-enhancing transgene in a population.

Domesticating Transgenes

Jonathan Gressel, a weed biologist in Israel, has proposed that one way to minimize the ecological effects of a transgene on the loose in a wild population is to make sure the transgene is integrated near another gene that would cause the transgene not to become fixed in populations.[19] Let's say we select a transgenic event in which the *deathstar* transgene is integrated next to a gene that prevents shattering. Farmers don't want crop seeds to shatter and fall to the ground either before or during mechanical harvesting. However, most weeds have adapted shattering as a trait to help spread their seeds in finding suitable germination and growth habitats. While the Bt or *deathstar* gene might be selected for, a *no-shatter* gene might be selected against in wild plants. Therefore, *deathstar/no-shatter* pair might be effectively neutral or even selected against in wild plants. And because they are closely linked on the same chromosome, they will not become quickly unpaired and therefore will be inherited together for a long time. If we could identify such a *no-shatter* gene, then we could make a new combination gene encoding a fusion protein, a new single protein that would have *deathstar* and *no-shatter* domains, so they would never become unlinked from one another. Progeny plants would always have both traits, and both traits would together control the fitness landscape of the recipient plant. I discuss the possible ecological effects further in light of genetics in chapter 10. In that chapter, I review some new exciting results that point to transgene domestication without purposefully linking transgenes to domestication genes. These results might change the paradigm of how we think of fitness-enhancing transgenes. Before I leave this issue for now, I want to reiterate that a transgene does not float in the genome by itself. The gene is attached to other genes that move from parent to progeny to progeny in a linkage group on a chromosome. This concept has direct applicability to gene flow and its consequences. A phenomenon called linkage disequilibrium should greatly ameliorate the selective advantage a transgene might confer during hybridization and introgression.

Round Round Get Around

I have presented a strategy to test the superweed candidates of canola in the United States, but a similar analysis could be performed for all crop–transgene combinations. There is also the issue of international borders. Pollen and seeds have a nasty habit of illegal migration—via truck, on the wind, and by animals. One aspect of regulatory control is lacking: transnational movement requirements and international biosafety coordination. For example, some people have pointed out that GM corn with no wild relatives in the United States that finds it way to Mexico could conceivably mate with teosinte, the progenitor of corn and a global treasure. Teosinte has its geographic center of diversity in Mexico and it exists nowhere else. While many GM crops will be commercialized at delimited regional scales, many GM crops such as corn could potentially be grown on all continents. Biosafety needs to be globalized as well. It is open to question whether GM traits have moved from illicitly grown corn in Mexico to traditional Mexican corn landraces (see chapter 5; figure 4.10). A priori, transgene movement within species is not as alarming as transgene movement between species, but, as we've seen, transgenes will likely not play a large role in affecting genetic diversity. Most people agree, however, that transgene movement should be monitored. It is a regulatory as well as a political problem when transgenic plants are illegally grown.

Aside from some superweed scenario, could we ever imagine any GM weed being significantly worse than *any* weed currently on the noxious weed list—even the tiniest bit? I think the answer is no, with the possible exception of something like transgenic johnsongrass (johnsongrass is already on the list). In this case, it would be prudent not to engineer sorghum for any transgenic trait, and, in fact, we should actively monitor for

Figure 4.10. Photos of teosinte, landraces, and elite corn. (Photos by Hugh Iltis, John Doebley, and Dennis West, respectively.)

any fitness-enhancing gene currently in sorghum moving to johnsongrass through hybridization. As we've seen earlier, we've learned much about non-GM plant breeding systems and ecology by the increased investigations of GM plant gene flow and ecology. This information will help us better understand the environmental impacts of agriculture.

Another Type of Superweed?

In chapter 8, the issue of insect resistance to insecticides and plant-produced insecticides such as Bt toxin proteins will be discussed. Simply, if the same insecticide is used over and over again, genetic changes in target insects may be selected for so as to render the insect resistant to the insecticide. The end result is that the insecticide will no longer control the insect pest and is rendered useless. A parallel situation may exist with herbicide overuse, which, by the way, has absolutely nothing to do with transgene flow from GM plants to non-GM plants. However, the evolution of herbicide-resistant biotypes might be accelerated and exacerbated by the habitual applications of a single herbicide. Roundup Ready crops do enable the use of only Roundup. Can a plant species that that is already a problematic weed become herbicide tolerant, so that it is no longer killed by spraying herbicide on it?

Millions of acres in the United States are planted every year with herbicide-tolerant soybean, corn, and cotton. There is a tremendous benefit to farmers for growing these plants. Weeds can be killed by spraying a single herbicide over the top of the crop; although the weeds are killed, the crop thrives because of the one or two added genes that render them tolerant to a herbicide such as Roundup (glyphosate). There is a twofold environmental benefit to this practice as well. The first is that glyphosate is a nontoxic chemical (except to plants) that degrades rapidly in the environment, which is in contrast with less environmentally-friendly herbicides. The second and even larger benefit to the environment is that farmers can better practice no-till farming. Farmers commonly remove weeds from fields by tilling the ground. In the absence of GM herbicide-tolerant crops, tillage reduces weed load so that chemical applications are kept at a minimum. The downside of tillage is increase in erosion of topsoil. In the past 10 years, the amount of no-till and reduced-tillage agriculture has increased more than threefold to 52 million acres in the United States. In 2002 Roundup Ready soybean varieties were grown on about the same acreage. Erosion has historically been a problem in places like the Mississippi River watershed, which includes western Tennessee. In western Tennessee, reduced-tillage agriculture has been practiced for many years to keep fertile and fragile soils from washing down the Mississippi to the Delta. And in no-till agriculture there are generally greater weed problems. Conservation tillage increases the importance of certain weeds such as horseweed, also known as mares tail.

It is important to note that not all plants are killed by Roundup herbicide; several plants are in fact tolerant. However, it was doubted that a glyphosate-resistant broadleaf weed such as horseweed would emerge anytime soon because of the herbicide's mode of action. But that is exactly what has happened. First described in Delaware in 2000, in the same year a few glyphosate-resistant horseweed plants were noticed in western Tennessee.[20] By the following year glyphosate-resistant horseweed was found on tens of thousands of acres, and in 2002, upwards of a half a million acres of crops in Tennessee alone contained Roundup-tolerant horseweed. Presumably, one or two genes inherent in the horseweed species allow it to survive Roundup applications, but prior to Roundup Ready crop cultivation, these resistance genes were very rare in the weed populations. The use of Roundup year after year, though, selected for the rare resistance gene(s), which spread rapidly across populations of horseweed, just as population genetics predicted it might. While this persisting situation is not good for farmers, Roundup-tolerant horseweed is not really a superweed. It can be controlled with other herbicides or tillage. It does, however, throw a monkey wrench in the works for Roundup Ready crops because its genesis was not anticipated, but it happened just the same. A Roundup Ready weed arose through mutation and selection, and is now a new weed problem for farmers. University of Tennessee scientists are studying the problem, along with others, to determine how this happened and to learn more about the weed's biology.[21] Independent from one another, my lab and scientists at Monsanto are attempting to elucidate the molecular mechanism of resistance. In addition, my lab is interested in the population genetics of the herbicide-resistant horseweed biotype. Ultimately, we need to know more about the genetic basis of weediness: weed genomics.[22] Thus, we will also have a new set of tools to dissect the molecular mechanisms of herbicide tolerance.

In this chapter, I have shot down practically all superweed candidates. The last example, horseweed, is not so quickly dismissed as the others. While it is controllable, the significant yearly population growth is exponential and overwhelming. For years, researchers interested in studying the risks of biotechnology have focused so intently on gene flow from transgenic crops, and now the evolution of herbicide-resistant biotypes has caught us off guard. Perhaps the scientific community's tunnel vision has betrayed us. While glyphosate-tolerant field mustard is on tens to hundreds of acres via transgene flow, glyphosate-resistant horseweed will soon be on millions of acres from spontaneous mutations and high selective pressure. And horseweed is not even one of the worst weeds. What's next?

The horseweed example shows that biology is complicated and that organisms will sometimes do the unexpected. We should not be so surprised. Does this uncertainty dictate that we ought to shut the door to biotechnology as a potential solution to real problems such as soil erosion? I don't think wholesale exclusion of technologies is the answer, but we have to realize that sometimes life in the real world goes awry, and contingency plans are needed. In addition, we must continue to develop scientific tools and well-trained people in the public sector who can mitigate difficulties when biotechnology wanders off the planned path.

As we'll see, it is often tricky to tell exactly if things are going wrong, or whether people only think (or hope) biotechnology has jumped the track. Sometimes it isn't obvious to the public how scientific results should be interpreted. On top of that, judgment errors can be made by scientists, science writers, and even science journals in publishing certain data. Unfortunately, science is sometimes hyped beyond the realm of its current level of understanding, and when data are not clear-cut to begin with, the public gets confused and might draw conclusions data don't warrant. Toss into the mix a contentious area such as agricultural biotechnology, and suddenly battle lines are drawn and war is declared.

5

Contamination

Transgenes in Mexican Corn?

The previous chapter focused on canola and gene flow to wild relatives, but plenty of research has examined other crops and their wild relatives. Although interspecific gene flow is of most interest because of the potential ramifications both inside and outside of agriculture, there are cases where transgene flow within a crop species is of interest. Still, it was surprising that the biggest gene flow paper of 2001 centered around corn. Toward the end of that year, I received a call from a science writer who had faxed me a paper that would appear in the prestigious journal *Nature* a few days later. The science writer wanted me to read the paper and then talk with him. Authored by assistant professor Ignacio Chapela and his graduate student David Quist, the paper claimed there had been introgression of the cauliflower mosaic virus 35S promoter into the genome of Mexican landraces of corn.[1] Furthermore, the corn they had sampled was in the state of Oaxaca, which is quite remote from where any transgenic corn had ever been legally grown (figure 5.1). In fact, the cultivation of GM crops had been illegal in Mexico since 1998. After quickly reading the paper, I spoke with the science writer the next day. Several things about the paper did not seem quite right, but I did not immediately follow up on them—activities associated with the end of the semester crowded my schedule. Little did I realize at the time that this 2001 paper would bring up the curtain on 2002's leading scientific soap opera.

Figure 5.1. Map showing the location of the U.S. corn belt, where most of the corn in America is grown, and Oaxaca, Mexico.

As the Corn Turns

In the November 2001 *Nature* paper, Quist and Chapela[1] claimed not only that there were transgenic DNA sequences present in traditional Mexican corn landraces, but also that transgenes had been introgressed into the native Mexican corn. The assertion of introgression was weighty, since it implied that transgenes had been stably integrated into the landrace genome, but the boldest conclusion of the paper was that a transgenic promoter, and possibly genes, were hopping around the genome of their newfound Mexican corn home. In the months that followed, many researchers in the field were astonished that these conjectures were published in the pages of *Nature*. After all, *Nature* is widely considered to be a premier scientific journal. The stark fact was that the data did not support the conclusions of the article. In fact, after I caught my breath during the semester break, I predicted the paper would be retracted,[2] and many other scientists in the field complained of the disintegration of the standards of the erstwhile queen of science journals. There was considerable fallout that ensued following the report.[3,4] More than 4 months later, *Nature* finally published two of the four submitted brief communications that revealed sizable fissures in the original data,[5,6] as well as Quist and Chapela's response to the criticisms.[7] Perhaps the biggest bomb was the unprecedented note from the editor, published along with the rebuttals, declaring that the paper should have never been published in the first place.[8]

In the next few pages, I dissect the controversy and examine the science behind the original paper and rebuttals while discussing the most appropriate molecular methodologies that should have been used. If illicit gene flow of transgenes has occurred in Mexican corn, what would be the ramifications? Would genetic diversity be decreased, irreversibly damaging Mexican corn landraces and agriculture? Furthermore, social and scientific questions of conflicts of interest were leveled during the controversy. How valid were they? Finally, a transgene containment strategy employing biotechnology will be explored.

Ignacio Chapela was, at the time, an assistant professor specializing in microbial ecology in the Division of Ecosystem Science, which is a section of the Department of Environmental Science, Policy, and Management at the University of California, Berkeley. David Quist was a Ph.D. student in the same department supervised by Chapela. Chapela's research group studied ecology and not molecular biology or biotechnology per se. Yet the results published in the *Nature* paper were entirely dependent on the execution of sophisticated molecular biology techniques. A perusal of Chapela's refereed publications indicated his team could successfully amplify DNA by using the polymerase chain reaction (PCR), which is a common tool now used by many non-molecular biologists. PCR is a sensitive technique used to amplify short spans of DNA using a DNA polymerase enzyme; after PCR is performed, the amplified DNA can be detected on a gel. To perform the analysis, one isolates DNA from an organism, performs PCR on it, and then separates the DNA products on a slab gel for visualization. While PCR is not technically demanding (many high school and college undergraduate biology labs perform DNA isolation from their own cheek cells and run PCR experiments to demonstrate DNA fingerprinting), the procedure can be prone to false positive results. This is especially true of the trickier types of PCR the authors used in the controversial paper: nested PCR and inverse PCR. A flawed experiment is apt to make the researcher think something has happened when in fact it has not, yielding a false positive. With the exception of the Chapela and Quist paper, I cannot remember a *Nature* paper being published solely on PCR data; more robust experiments are expected before a paper is typically deemed publishable in such a journal. Nonetheless, it is clear now that the appropriate experiments were never performed.

Burnt Bridges

It is notable that Chapela is perhaps most infamous for his outspoken opposition of the 1998 Novartis (now Syngenta)–UC Berkeley deal that provided up to $25 million over 5 years to the Department of Plant and Microbial Biology. Novartis agreed to fund research in that department in exchange for first right of refusal to license appropriate inventions from affected faculty labs. In contrast with Chapela's own department, the Department of Plant and Microbial Biology contains several world-class plant molecular biologists. From the notes and acknowledgments in the original paper, it does not appear as if Quist and Chapela ever sought to tap into the pool of molecular expertise resident in the plant and microbial biology department in their study. For that matter, no molecular biologists in the Berkeley area seem to have been involved in the study, which I find especially curious. I venture to guess that the inclusion of molecular biologists as authors or collaborators on the *Nature* paper would have greatly aided the technical aspects of the experiments.

The *Nature* Paper

To move on the main questions at hand, why was the paper finally retracted and what were the methodological flaws that led to its retraction? Hindsight shows that the paper should have never been published in the first place. The methods used were inappropriate, a fact recognizable by graduate students in molecular biology. The authors claimed to have detected, in Mexican landrace corn, the cauliflower mosaic virus 35S promoter, a frequently used promoter in transgenic plants. It would have been much more straightforward to detect the transgenes themselves. In the case of the 35S promoter, small quantities of virus in the environment (field or lab) could have caused a false positive result. Assaying directly for insect-resistance or herbicide-tolerance transgenes would have greatly eliminated the possibility of false positives from stray contamination. These types of transgenes are in American fields in large numbers—millions of acres every year. Corn varieties containing these transgenes would have been the likely culprits cultivated illegally in Mexico. The United States would have been the ultimate source of transgenes potentially flowing into Mexican landrace corn. Although PCR might have been used to test for the presence of Bt *cry1Ab* or Roundup Ready genes, Southern blot analysis is a more certain and conservative analysis for testing whether the transgenes themselves had been integrated in the corn genome, a fact a molecular biologist working in the field would have recognized. In this tried-and-true procedure, the DNA of interest is run on a gel and then transferred it to a nylon membrane for probing. Transgene probe DNA is used to directly detect the target sequence on the membrane. One can determine if the transgene is present and intact by its expected size. After being criticized by other scientists,[5,6] Quist and Chapela finally did perform a DNA dot blot assay, which, unlike the Southern blot, gives no indication of molecular size of insert, copy number, or even if the signal observed is from a transgene that is actually integrated into its host's genome.[7] The latter point of transgene integration and introgression was hotly contested. Contamination of PCR was a continual question in the minds of critics.

Nature Publishes the Criticisms

Two scientific communications were published by the journal that sufficiently and convincingly rebutted the original paper. The first claim of transgenic promoter introgression into Mexican corn landraces was found to be implausible from the original data. Quist and Chapela used the ultrasensitive nested PCR to detect the presence of the 35S promoter in pooled samples of ears of corn. They claimed to have demonstrated that few kernels on any given ear (i.e., individuals in a population) were transgenic. Nested PCR amplifies DNA and then amplifies it again in another round of experiments. As pointed out by Metz and Fütterer,[5] a few transgenic kernels on ears of corn indicated F_1 hybridization had occurred, but certainly not introgression, which would have required repeated backcrosses and the resulting stabilization of the transgene in the new host genome. Most of the kernels on a given ear should have been transgenic if the corn plant bearing the ear had been introgressed with transgenic sequences. To my knowledge, no data have ever been published showing that unintended transgenic DNA has been introgressed into corn landraces anywhere or, for that matter, into any unintended corn genome. If strong selection for the transgene were present, one would expect introgression might occur quite rapidly in an intraspecific cross, and this simply hasn't happened. There would be absolutely no genomic barriers for a corn × corn cross. Given that the transgenes in question would have been subject to positive selection and no introgression has occurred, then dire worries of transgenic decreases of genetic diversity seem unwarranted.

The second claim of Quist and Chapela was that the 35S promoter sequences (and presumably those from associated genes) were moving around the genome of Mexican landrace corn. Their original data came from inverse PCR (i-PCR), in which host DNA is experimentally circularized and known sequences are used to prime outward for DNA amplification around the circle. The power of i-PCR is the ability to obtain the identity of unknown flanking sequences from the utilization of known sequences. Both groups who criticized the 2001 *Nature* article, Metz and Fütterer[5] and Kaplinsky et al.,[6] convincingly demonstrated that the original i-PCRs were flawed. One would expect transgene sequences from Bt or glyphosate tolerance genes to be adjacent to putative 35S promoter sequences, but none was found from the i-PCRs. Quist and Chapela also seemingly misinterpreted sequence similarity between a gene sequence that has been used in plant transformation plasmids (an *adh1* intron) for genes that are naturally present in all corn genomes (the *adh1* gene and a *bronze1* sequences).[6] This same group of critics also pointed out that the i-PCRs were not technically performed correctly, which would also lead to artifactual results. PCR can be tricky. Nested and inverse PCR can be extremely tricky procedures.

If, indeed, the Berkeley scientists had been correct in their conclusions that promoter sequences and transgenes were genomically unstable, it would be an extremely significant finding. Such a finding would lead to a paradigm shift in plant biology indicating that transgenic sequences introduced via biotechnology behave totally unlike natural gene sequences. Scientists would be scrambling to discover the mechanism of this newfound genetic instability in the face of the 38 trillion genetically modified plants grown in the United States containing this same 35S promoter. One would wonder how such a giant effect had dodged the collective watchful eye of science and why no commensurate ecological disaster had yet ensued.

Because of the now-obvious flaws, the editor of *Nature* stated his regrets over publishing the article in an unprecedented action that distinctly distanced the journal from the article.[8] Still, the authors stood by their findings (at least the ones they still believed to be valid).

Consequences of Gene Flow to Mexican Corn Landraces

So, what parts, if any, of the Chapela and Quist results are convincing, and what are their ramifications? Even the critics agree that it is certainly possible that transgenic corn could have been illegally grown in Mexico, and that transgenes are flowing among varieties and landraces of corn.[3–6] There would be no interspecific genetic barrier, and food containing transgenic corn from the United States is shipped to Mexico every day. Mexican migrant farmers in the United States could easily have purchased corn seed for cultivation and shipped it to Mexico. There are many opportunities for Mexicans to obtain GM seeds. Indeed, an April 19, 2002, newspaper report cited Mexican government scientists' claim that transgenic DNA is present in corn gene banks in Mexico.[9] No paper has been published, but I do not doubt that there is a limited amount of transgene flow in and out of Mexican landraces. So what?

What exactly would be the ecological and agricultural ramifications of the presence of transgenes in landrace corn populations in Mexico? In the previous chapter, teosinte, the progenitor of corn, was discussed. There is no evidence to suggest that GM corn has interbred with teosinte, but that could be the next big story. Even if it is the next big story, there is no empirical evidence to suggest that it would be different than non-GM corn breeding with it. There are no convincing data, however, showing that corn can introgress its genes into teosinte in any significant amounts. In fact, scientists have shown that corn probably arose from a single domestication event from teosinte and that uni- or bidirectional gene flow is absent (crop-to-weed introgression, reviewed in Stewart et al.[10]). Right now, one concern of many people is whether the south-of-the-border transgenes in the geographic center of diversity of corn will decrease genetic variation found within Mexican landraces.

Even though corn is, for all intents and purposes, a man-made entity that is dependent on humans for its survival, it is important to maintain its genetic diversity for future crop improvements. There are three mutually exclusive scenarios that could play out if transgenes ever were to become introgressed into traditional corn landraces: genetic diversity could decrease, stay the same, or even increase. Even if a transgene were highly selected for by humans, only the immediately linked genes would accompany the transgene in progeny; it would follow the rules of population genetics described earlier. There would be a degree of initial genetic variation that would be maintained. The transgene would act as any other gene in corn that has undergone the same process for millennia. Although more research needs to be performed to assess the consequences of the adventitious escape of transgenes into plants in geographic centers of diversity, there is no cause to jump to the conclusion that the effect would be even slightly negative. A transgene could actually increase genetic diversity by increasing the effective population size of interbreeding corn plants. A greater population size would allow greater recombination and mutation and, hence, would allow greater genetic diversity to be maintained. If there is an effect (positive or negative), it will be confined to possibly affecting subsequent genetic improvement of the crop. There would be no effects in nature. Ramifications of this issue are more political than scientific.

And on the political end of the scale, are any U.S. entities (governmental or corporate) liable for transgene movement? Should U.S. regulatory agencies deregulate a transgenic crop in the United States assuming that it will not be illegally grown in other countries? This last assumption is naïve. It has been documented that illegal transgenic soybean has been grown in Brazil, and contraband GM cotton has been harvested in India.[5] Unlike the scientific picture, which is rather clear, policies and politics are muddy and emotional. Unfortunately, science-based regulation is not ubiquitous throughout the world. In Europe, a unified, purely science-based regulatory framework has not prevailed in the montage of varied political points of view among European Union nations. As mentioned before, a global biosafety view of transgenic plants and, in fact, all crop varieties, is needed, because plants grow, reproduce, and move around without respect to boundary lines. Any attempt of creating such a framework now would most likely fall into the same trap as in Europe, unless some serious constraint to make the regulations science based were forced upon the system. An organization called Croplife International (www.croplife.org) is investigating regulatory and standardization protocols surrounding the biosafety issues of globalization of plant biotechnology.

Objectivity and Conflicts of Interest

A second, and rather surprising, feature of the Quist and Chapela *Nature* debacle spotlighted the ulterior motives and objectivity of myriad scientists, thus potentially decreasing the credibility of science in public perception. Just when I thought the dust had settled on the Mexican standoff—with the arguments of science winning the day, another round of letters was sent to *Nature*. This time, the griping parties were Quist and Chapela's allies, who felt their colleagues had not been treated fairly by the scientific community or by *Nature*.[11,12] These authors did not approve of the fact that *Nature* retracted the paper without Quist and Chapela's consent, and they especially did not like that some of the scientists who criticized the 2001 paper had received industry funding. This latest round of correspondents was affiliated with Chapela's academic unit at Berkeley and they were also disgruntled by the fact that (Novartis) Syngenta had supported Berkeley in the strategic alliance of 1998. They felt that the scientists criticizing Chapela and Quist were biased because of their own industry funding for research.

Quist and Chapela's allies argued that their colleagues were vehemently disputed because they presented unexpected data that ran counter to engrained scientific dogma. In the words of Worthy et al., "In such an environment, it is difficult to imagine fair and equal consideration being given to work that challenges the commercially vested interests of ag-biotech and the assumptions of reductionist molecular biology. Quist/Chapela obviously represented such a challenge. That fact—not the quality of their work—together with the politics of university-industry relations, remains central to their paper's troubled reception."[12] These allies called into question the motivations and funding of the original critics and said that criticisms and the resulting retraction of the original paper were not based on science at all. As Metz and Fütterer (original critics) pointed out, "Our concern was exclusively over the quality of the scientific data and conclusions, which would have been the same whatever the motivation of the criticism"[13] (p. 187). And Nick Kaplinsky, one of the critics of the 2001 paper, echoed this case, saying that the correspondence "was a critique of poorly conducted and interpreted science and was not pro- or anti-GMO or industry."[14] Dr. Kaplinsky also made the point that there is a seeming double standard with regard to biases. Chapela is a member of the board of directors for PANNA, the Pesticide Action Network of North American, which is a vocal opponent of genetic modification.[14] The PANNA Website states that Chapela used to be an employee of Sandoz (which merged with Ciba to form Novartis, now Syngenta), and he "renounced corporate exploitation of genetic resources . . . and now researches and teaches about these issues while assisting Mexican indigenous organizations, NGOs and others to meet challenges related to genetic engineering."[15] In the end, I think science won over mudslinging.

In my opinion, the second, nonscientific group of correspondences were us-versus-them exploitations that had no place in a scientific journal. From the November 2001 publication of the original paper, the main issue was always the science, and there have been no follow-up papers. Can science and technology mitigate transgene flow or even prevent it from happening all together? Is it possible to keep transgenes corralled into intended agricultural fields to prevent their wanderings?

Terminator Technology

Three years earlier, in 1998, the plant biotech controversy of the year was an invention by USDA and Delta and Pineland scientists soon dubbed "Terminator." The inventors of the technology, led by Mel Oliver at the USDA Agricultural Research Service (ARS) and a group at Delta and Pineland, Inc., originally named their invention the "technology protection system," but the vision of Arnold Schwarzenegger coming back to wipe out plant life guaranteed that the invention would be popularly known as Terminator technology.

News broke about Terminator (not to be confused with the terminator that is used in gene constructs [chapter 3]) when a U.S. patent was issued entitled "Control of plant gene expression."[16] RAFI (now ETC Group) very cleverly hung the devastating tag "Terminator" on the invention and was successful in propagating fears that poor farmers would be devastated by corporate, biotechnology-driven agricultural companies epitomized by Monsanto.

The crux of the technology was to control gene flow by killing embryos in seeds using a molecular gene expression system that could be controlled by the party that sells seed. The technology could indeed protect a company from losing its investment in the development of a transgenic plant by preventing a farmer from growing saved seed from year to year instead of annually buying transgenic seeds from the company that developed them. The agricultural industry perspective on this topic is similar to that found in the entertainment and publishing industries, which strive to protect investments from illegal copying of films, music, and books. It is reasonable to expect that the individuals who own intellectual property or copyrights should stand to gain fair compensation from their work. That said, the scientifically interesting aspect of Terminator technology is in limiting gene flow from transgenic to wild plants and plants of the same species. I'd rather think of Terminator as an environmental protection tool in need of a new name. Terminator and similar technologies would be a great boon to minimizing the ecological risk of unwanted hybridization and introgression. In the case of Mexican corn, shipments of corn seed to Mexico could be rendered useful for food but unable to germinate to use as a crop, thanks to Terminator.

The mechanism of Terminator lies in the precise control of at least two different gene constructs (figure 5.2). Immature embryos in seeds can be

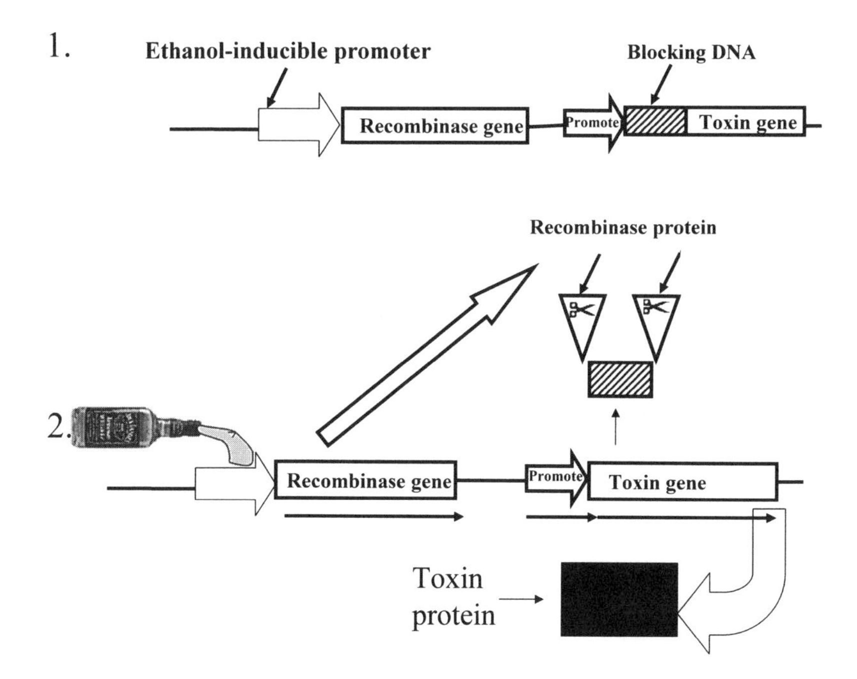

Figure 5.2. One example of how a terminator mechanism would operate to control seed germination ability. The genes would be in a transgenic plant and the cascade activated by ethanol application. (1) A recombinase gene is under the control of an ethanol inducible promoter. In this case no ethanol is applied. Result—toxin gene is not expressed since blocker DNA remains in place and seeds can germinate. (2) Ethanol is applied and turns on expression of recombinase gene. The recombinase acts to remove the blocking DNA from the toxin gene. Result—toxin gene is expressed and kills embryo in seeds so they cannot germinate.

killed if a toxin gene, let's say one coding for a diphtheria toxin, is expressed only in the developing embryo. If it were to be expressed in the rest of the plant, it would kill the host before it had the chance to produce seed, so a late embryo-specific promoter must be used to control expression of the toxin gene. Between the toxin gene and the promoter is a blocking DNA sequence. This sequence acts as a spacer to prevent the toxin gene from being expressed. The blocker can be engineered so that it can be excised by a recombinase protein, so precise inducible control of a recombinase gene is required. That inducer might be a chemical that normal plants are not typically exposed to, such as ethanol. When the company is producing seed lots to sell to farmers, the blocker should stay in place. The recombinase gene would not be induced (no alcohol applications on the "original" seeds), and the seeds would be planted and grown to yield the seeds to sell to farmers. Before selling farmers these second-generation seeds, they will be sprayed with alcohol to activate expression of the toxin gene. Since the embryo is already mature, the toxin gene will not be expressed in the seeds sold, but as the seeds are germinated, the blocker will be removed automatically from all cells. When it comes time for seed set and embryo development, the toxin will be produced to kill the developing embryo, while leaving the protein and oil in place for the farmer to sell—all contained within the "dead" seed. There are several details that would have to be ironed out for such a system to be practical. The most obvious one is that the toxin would have to be proven nontoxic to the humans ingesting the seeds in foodstuffs. As an ecological application, the technology would have the desired effect of rendering all the seeds sterile that resulted from pollinations outside the field of interest; the crop field. The system should not be leaky, thus not allowing escapes to occur.

The technology has not been fully developed, so it is still too early to get excited about either a beneficial ecological effect or upset that the technology might harm poor farmers. However, it is too bad that the technology garnered such bad press and that skeptics generally squashed it before it could be proven. RAFI and others painted such a grim and vivid picture of Third World farmers as slaves to multinational conglomerates that Terminator never really got a fair hearing. Never mind that poor farmers do not buy seeds from Monsanto. They still save their seeds from their previous year's crops, which came from their previous crops, which are not transgenic. Nor would Terminator really change the paradigm for farmers in the developed world who normally purchase seeds of the latest and best-yielding varieties each year. Furthermore, these varieties are often hybrid seed; that they could not save them and grow them the following year, even if they wanted to. If they tried, they would no longer have the pure-breeding hybrid seed. In spite of the original objections Terminator garnered, it is time that companies and other biologists investigated technologies that will decrease transgene flow from crops to wild relatives and landraces, be it Terminator or some derivative. Ecological reasons for doing so are compelling.[17] But there is still a need for an eco-friendly name. These days, the generic name for Terminator and related technologies is gene use restriction technology (GURT).

In addition to GURTs, there are a number of other technologies that might corral transgenes inside their intended hosts. Recombinase technology might be used to excise transgenes from certain cells. Plants can be engineered to be male sterile, either using recombinase or other technologies so that transgenic pollen cannot pollinate non-GM plants. Other scientists suggest that engineering the chloroplast instead of the nuclear genome might keep transgenes from escaping through pollen (pollen grains don't have chloroplasts). There are combinations of technologies that might prevent transgenes from being introgressed into wild relatives, or even in crops where they are not wanted.[10]

Before the 2001–2002 controversy, the king of all biotechnology controversies featured the monarch butterfly as star and GM corn as villain. After a brief lesson on the development of plant biotechnology as industry, I focus on the science underlying the press and alarm that ensued following the publication of another *Nature* article.

6

Killer Corn

Monarch Butterfly Exterminators?

Biotechnologists as Environmentalists

Once upon a time, environmentalists at Monsanto dreamed of the day chemical insecticides would be a thing of the past. Indeed, in the 1980s, young, brash scientists foresaw methods to engineer crops to produce their own environmentally safe insecticides, rendering chemical insecticide sprays unnecessary. Their reverie featured an elegant biological solution for pest control instead of the conventional repetitive chemical applications that then dominated the crop protection landscape. They envisioned high specificity against certain pests to avoid the downside of broad-spectrum insecticides. Only the crop-damaging insects would be killed; the beneficial insects and other animals and microbes that occupy farmers' fields would be unharmed. No one, not even those in the agrochemical industry, believed that synthetic insecticides were wholly environmentally benign. Nontarget effects in the field were documented. While protecting the crop from target pests, beneficial insects were sometimes inadvertently killed. In addition, there were human health hazards that ranged from ingesting pesticide residues on food items to dangers from overspray and the poisoning of farm workers. Finally, there were larger ecosystem effects as chemicals made their way through food chains and into waterways and groundwater. These facts did not, however, prevent the agricultural industry from generally disagreeing with Rachel Carson. She had convincingly made the case that insecticides would destroy wildlife and, ultimately, humans. To her, the risks far outweighed the benefits. Industry scientists, however, all but admitted that chemicals were a risky and necessary evil needed to effectively protect farmers' crops from insect pests. The ends justified the means. Not much had changed until biotechnology appeared on the scene, but Rachel Carson predicted the arrival transgenic plants more than 20 years before their first production.

In *Silent Spring,* her seminal book written in 1962, Rachel Carson painted a very nihilistic view of the world in which agricultural chemicals were the mainstay for crop protection.[1] In the backdrop of the early 1960s, DDT's shadow was still extremely dark and foreboding on the consciousness of the public. Carson successfully employed the DDT disaster as the lightning rod in her book. Twenty years before the book's publication, the miracle insecticide, DDT, had proven to be a lifesaving chemical—a penicillin of pesticides. It effectively quenched the spread of malaria by controlling its vector, mosquitoes. Unfortunately, DDT was also a resilient compound that remained intact in food chains, able to move through food chains from plants to insects to birds, and then to humans. The intact chemical compromised the ability of birds to reproduce, hence resulting in the prophetic "silent spring" with the dearth of songbird populations wiped out by DDT. Just as Paul Ehrlich would do just 6 years later, Rachel Carson projected the grim future of a tattered natural world.[2] While Ehrlich wrongly predicted the development of an overcrowded world in which mass starvation and rampant disease would destroy humanity and the world's ecosystems, Carson was the prognosticator of an agricultural-caused ecological disaster. Although both doomsday scenarios were proven to be false, they did provide the inspiration for a new social consciousness and protest industry: the environmental movement, which will be examined more extensively in chapter 11. The environmental movement of the 1970s not only influenced the general populace, but also impacted scientists, including the new breed of biotechnologists. They wanted a better, greener world as much as anyone and saw biotechnology as part of the solution.

In *Lords of the Harvest,* Daniel Charles describes the mindset of the young, environmentally conscious, industrial biotechnologists working in Monsanto and other companies in the 1980s.[3] These scientists, who had the knowledge and the newfound tools to produce transgenic plants, envisioned a transformed agricultural world. In their world, farmers would no longer solely rely on synthetic chemicals to protect crops. In the new world of agriculture, biologists would be the heroes. The new biologist, the biotechnologist, could engineer the crop to produce its own pesticide without the aid of the chemist. Indeed, the chemists at Monsanto were not entirely thrilled at the epiphany that the biological upstarts conjured. But, as they say, the rest is history, as the new biotechnologists in the 1980s were successful in their quests to change the industry and agriculture itself.

The new biotechnologists agreed with Carson that biological solutions were better than chemical solutions. In Carson's words,

> A truly extraordinary variety of alternatives to the chemical control of insects is available. Some are already in use and have achieved brilliant success. Others are in the stage of laboratory testing. Still others are little more than ideas in the minds of imaginative scientists, waiting for the opportunity to put them to the test. All have this in common: they are *biological* solutions . . . entomologists, pathologists, geneticists, physiologists, biochemists, ecologists—all pouring their knowledge and their creative inspirations into the formation of a new science of biotic control.[1(278)]

This quote represents the epitome of what Monsanto's biotechnology effort embodied in the 1980s through today. And while Rachel Carson did not know much about molecular genetics (although she did know about and was an advocate of using *Bacillus thuringiensis*, the bacterium that controls many types of insects and has been the source of the many Bt transgenes used to control insects in GM plants), biotechnologists were confident that genetics and biotechnology were the surefire biological solutions to insect pest problems. It was unconscionable to them that anyone could object to biology over chemistry—especially environmental activists. It was foreseen that the environmental groups would embrace the new agriculture with open arms, since biotechnology then was the seeming consummation of Rachel Carson's prophecy. Of course, in hindsight we know that green groups sorely objected to the new biotechnology and its implementation. While it is obvious to me that *Silent Spring* was the forerunner of biotechnology, several of Carson's disciples have berated me for suggesting this.

Bacillus thuringiensis

The obvious first choice for a gene that could be used as the new transgenic plant insecticide was one from *Bacillus thuringiensis* (Bt). It was envisaged to be a useful tool for the control of caterpillars on several crops. In addition, it was also known that there were different Bt strains (and hence, various genes and proteins) that were effective against all kinds of insects—not just lepidopterans (butterfly and moth caterpillars), but coleopterans (beetles) and dipterans (flies and mosquitoes), among others. That Bt had been used by organic farmers and was safe was a tremendous boon to biotechnologists, since Bt transgenic crops represented a large opportunity for a successful and safe first biotech product. There were several attractive aspects to plant-produced Bt proteins for insect control. First, people and wildlife had been exposed to Bt for many years, and its safety was generally and universally accepted. Second, it appeared that the plant could express a single Bt endotoxin gene, and the plant-produced Bt protein would kill target insects. The single-gene solution spelled a relatively simple biotechnology answer to an otherwise complicated problem of how to kill target insects without harming nontarget species. Third, each year farmers lose significant portions of their crops to insect damage, mainly from caterpillars and beetles. Nowhere was it worse than in cotton and potato crops. These farmers would embrace an effective and safe solution to insect problems that did not rely on toxic chemicals.

The infamous boll weevil (a coleopteran) had dethroned cotton as king in the southern United States in the early part of the twentieth century, causing farmers to diversify the kinds of crops they planted (figure 6.1). Since then, several lepidopteran pests, such as bollworm, had plagued cotton cultivation. The availability of chemical insecticides in the middle part of the twentieth century was a terrific boon to these farmers, allowing them to expect a consistent and reasonable cotton harvest from year to year. But it came at a cost. The farmers would have to pay crop dusters to spray their fields with several insecticides multiple times per growing season.

A similar situation existed for potato, which normally suffers from heavy feeding pressure from the Colorado potato beetle. Even chemical insecticides have not been widely successful in controlling this insect. Some farmers go as far as attempting to control Colorado potato beetles via flamethrowers mounted behind a tractor.

Since cotton had higher value and more acreage than did potato, Bt cotton was produced first. In addition, the two genes that would be effective for cotton (*cry1Ab* and *cry1Ac*) could also control European corn borer, which is sometimes a devastating pest on corn. Compared to cotton and potato, insect problems in corn are minuscule, but corn is the biggest crop in the United States. European corn borers make their homes

Figure 6.1. Photo of the Boll Weevil Monument in Enterprise, Alabama. Note the boll weevil at the statue's pinnacle. (Photo by Ronald Smith.)

inside corn stems where conventional pesticides cannot reach. So transgenic Bt corn seemed like a product that farmers would buy, even if it did not make a big difference to their bottom line every single year. Farmers would view Bt corn as an insurance policy against relatively infrequent but severe infestations of the European corn borer. In addition, Bt genes would be effective in another occasional pest, the corn earworm. When I was a boy, I was typically the one chosen to shuck the sweet corn before dinner on a summer's eve. I grew to despise corn earworms for the very gross surprise when they were exposed upon shucking: Not only did they destroy kernels on the ear and leave a very nasty brown residue, but also they would rear their heads back defiantly as if to say "you're next, kid." Indeed, this could be the Freudian impulse behind my research with insecticidal plants—to finally get even with the menacing corn earworms.

In 2002, about half the cotton crop and approximately one-third of the corn crop in the United States was Bt transgenic. Even though Bt transgenic potato was widely adopted by farmers in the mid- to late 1990s and was effective in controlling Colorado potato beetle, the McDonald's hamburger chain ban on GM potatoes effectively destroyed the market for Bt potato, the only GM potato on the market. After 2000 the Bt potato was history. Who could have predicted that a fast food chain could single-handedly kill a nutritionally and environmentally safe product? To see why Bt is safe to eat and why it is deadly for specific insects, we must understand the mode of action.

How Bt Works

The bacterium *Bacillus thuringiensis* is one of the closest relatives of *Bacillus anthracis*, the bacterium that causes anthrax. Like the anthrax bacterium, it lives in the soil, forms spores, and produces crystal toxins (see figure 2.1). Unlike those produced by anthrax, Bt toxins affect the digestion of a narrow scope of insect types and are harmless to humans. Bt seems to live to kill insects and reproduce in their bodies. It is a completely natural insecticide, just as the bacterium *Agrobacterium tumefaciens* is a completely natural plant genetic engineer. The mode of action is the key behind Bt's desirability. While scientists have been aware of the high specificity of toxicity for a long time, the precise mechanism of how Bt endotoxins kill insects is still being unraveled. The main requirement for susceptibility is the presence of Bt endotoxin-binding receptors in insects' midgut cells. If the receptors happen to be missing, then the insects will be resistant to Bt's killing power. Why insects have these receptors in the first place is somewhat of a mystery, but it is clear that they are required for endotoxin binding. We know that after endotoxins bind, they poke holes in gut cells, and the insect dies from losing water and food.

Bt toxins are first synthesized as protoxins. Bt protoxin proteins are very large (molecular weight of around 130,000 Daltons; the molecular weight of table salt is 43 Daltons). Protoxins are digested by the host insect, which chops them to about half of their original size. It is these chopped toxins that are active and deadly. Truncated versions of Bt endotoxin genes, when expressed in GM plants, render this digestion unnecessary. Just like any protein, the function of Bt endotoxin is determined by its three-dimensional shape, which, in turn, is determined by the amino acid sequence. The activated Bt toxin protein has three parts; all the parts (or domains) are needed for action (see figure 6.2). Domains II and III are important for toxin binding to midgut cell receptors. The amino acids in these domains match up and bind with the insect's receptors. If there is insufficient alignment on the part of the Bt toxin, or the receptors change via mutation, then the toxin will not harm the insect. Once the Bt toxin protein binds to the surface of insect gut cells, then domain I creates channels in gut cells. The three-dimensional shape of domain I can be likened to series of corkscrews (figure 6.2). These corkscrews insert into midgut epithelial cell membranes, which rapidly disrupts the cellular balance of the insect's digestion process (figure 6.3). Insects die within hours of ingesting Bt. Domain I seems to be similar among most Bt toxins, while domains II and III are more variable among different kinds of Bt toxins. So insect-killing specificity is determined by the variation in domains II and III.

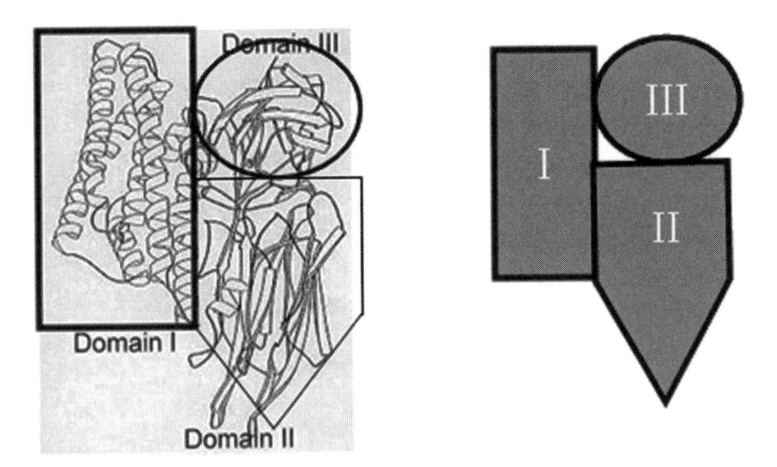

Figure 6.2. (Left) A protein ribbon diagram of a Bt endotoxin as represented from crystallography data, showing the three functional domains. (Right) A simplified block version.

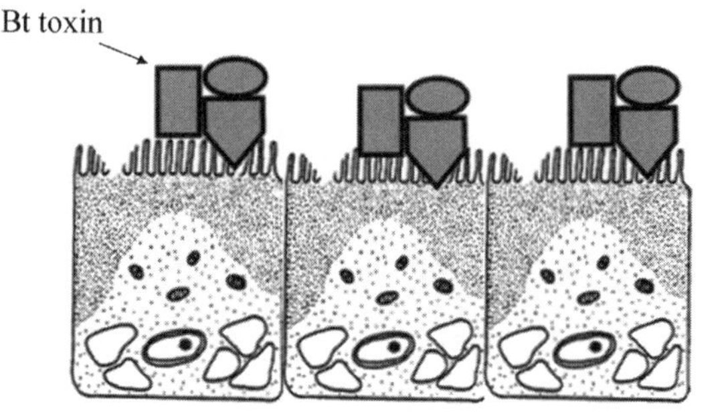

Insect midgut cells that have bound Bt toxin.

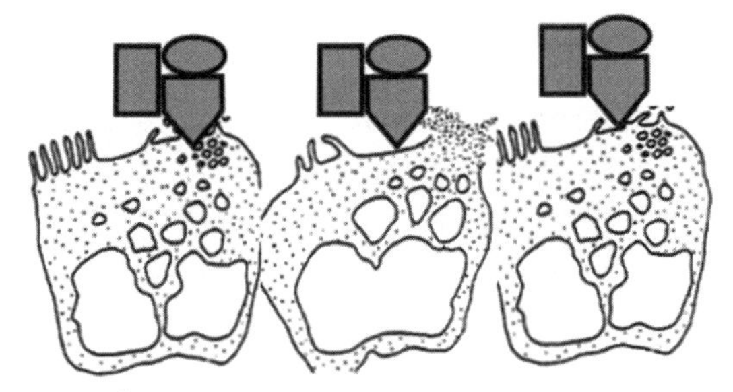

Same gut cells a few hours later– note the damage and leakage.

Figure 6.3. Bt toxins bind to insect midgut cells. Beginning a few hours after binding, the Bt toxins begin to disrupt osmotic processes in the cells. The damage kills susceptible insects very quickly.

Understanding the underlying mechanisms of the specificity of Bt toxin binding and its subsequent toxicity is vital to grasping all the potential environmental ramifications of Bt that will be discussed in this and the next two chapters. From here on, a certain kind of Bt toxin will be our topic: the type that has been used in plants for the purpose of killing caterpillars that feed on corn and cotton. This type has two variants. The Bt Cry1Ab toxin has been engineered primarily in corn, and the Bt Cry1Ac toxin in cotton. "Cry" stands for crystal endotoxin, and "1Ab," and so forth, is the systematic designation for the various families and kinds of individual toxins, which is related to toxin specificities. As the naming of these two toxins suggest, Cry1Ab and Cry1Ac are closely related. Of the thousands of variants of Bt toxin proteins that exist in nature (each killing a specific insect host), the two that are principally used in agriculture are similar, but vary enough that they optimally affect different species of caterpillars. Unlike most chemical insecticides that target the neurotransmitters that are ubiquitously prevalent in all insects (and many noninsects as well), Bt proteins only kill insects that have receptors that match the variations in Bt endotoxin domains II and III. So it is not surprising that many biologists are busy studying the genetics and physiology of the midgut cellular receptors of various kinds of insects to better understand why some insects are killed by Bt endotoxin proteins and why some are unaffected. They are interested in the tremendous genetic diversity of receptor binding and host specificity, but the research is also important for addressing nontarget toxicity questions—the insect side of the equation. Other biologists are studying the Bt toxin protein side of the equation. These scientists compare various endotoxin amino acid sequences in domains II and III and ask how this variation correlates with specific toxicity. Companies, especially, are keen on isolating new toxin proteins (and, hence, genes) so GM plants can be produced that have specific toxicities to a wide range of insects and crops. Most of the readers of this book have probably not heard or read anything in the popular media about any of the fascinating and timely biological research on insect toxin receptors or the biodiversity of Bt toxins. Instead, certain ecological studies have been widely publicized and hyped. It is important to move past the headlines to critically examine the larger ecological effects of Bt transgenic plants in and around the farm.

Bt Corn Pollen and the Monarch Butterfly

Most people can vividly recall their exact situations in life when they first heard certain traumatic news events. For example, I was 3 years old when President John Kennedy was shot and I clearly remember standing in my kitchen when my mother's eyes welled up with tears. The typical day's activities stopped as I watched her watch the sad scenes of the day unfold on a small black-and-white Philco television set. Then, 14 years later indelible memories were made when the radio newsman interrupted my joyride by announcing that Elvis Presley was dead.

With the same enhanced awareness I also vividly remember first hearing the story about the day the monarch butterfly died, about John Losey, the assistant professor at Cornell, and the villain: Bt corn pollen that killed poor pitiful monarch butterfly larvae. It was a warm late-spring day in 1999, and I had ironically just arrived for a lunchtime meeting with a group of scientists at Syngenta in Research Triangle Park, North Carolina. As the three corporate folks and I were stuck in I-40 traffic en route to a restaurant that used to be a tobacco factory, I wondered why everyone in the Cadillac seemed a little glum. They then broke the news to me. *Nature* had just published a short article written by a team of Cornell University researchers. Within a matter of hours news had spread: genetically modified corn may be hazardous to monarch butterflies. As I recall, at the time, none of us had yet read the actual scientific report that had been published that day in the journal, but the outlook from my corporate friends was gloomy.

The monarch debacle came on the heels of another bad press event. Only a few months earlier, Arpad Pusztai, a scientist with the Rowett Research Institute in Scotland, had gone on British television to disclose how his research had shown that GM potatoes caused nutritional and health problems in rats. He had performed a feeding study and publicized the results even before he had written a scientific paper. When Pusztai's paper was finally published a year later, it was heavily criticized, and it was dispatched by most scientists as a flawed study.[4] But the scientific consensus did not really turn around the negative public perception of biotechnology that the initial television interview had produced. Agricultural companies such as Syngenta (Novartis at the time) were forced into a defensive position and were spending copious amounts of time and money reassuring the public that their products had gone through thorough regulatory scrutiny and were safe for consumption (Pusztai's work was on potato engineered for a lectin gene, a transgenic potato that no company was going to put on the market anyway). Indeed, the Pusztai interview was the event that many people had regarded as the catalyst for extreme negativity in Europe with regard to GM food. So it seemed to my industry hosts that their company would have to fend off more negative publicity, assume the position, and take their licks. It was a real bummer of a lunch that day, even if I was not paying. It ended quickly and I was back on the road to the university with the top down, daydreaming of Elvis and the future of plant biotechnology. As far as any of us could see, there were only two redeemable things about this situation, compared with the Pusztai event: it was not on TV and it was not about food.

First, there was an actual scientific paper to accompany all the media attention, so there would be scientific data to address, and refute, if necessary. A journal editor typically arranges for two or three other scientists in the field to review a submitted paper, and it receives anonymous critiques before it is published. The reviewers and editor ensure that the data are sound and novel and that the paper is worthy of publication in a scientific journal. Sometimes a paper is rejected. When a paper is published, the university employing the authors might do a press release announcing the paper; universities are always hungry for good public relations, and a timely study in a prestigious journal is fuel for a press release. Then we are off to the races. Sometimes this process gets played backward like the Pusztai ordeal or if universities foolishly release scientific stories before the publication of a refereed paper. In these cases, there is no hard evidence for refutation or even discussion, since there are no available data to discuss. A media feeding frenzy can easily ensue with nearly no scientific information, such as when human cloning hit the news in early 2003 with absolutely no scientific data.[5]

The second advantage that the Losey situation had over the Pusztai debacle (an advantage if one is concerned that biotechnology gets a fair hearing) is that the subject was environmental effects rather than food safety. Even the most fanatical environmentalist is more concerned with his personal health than the health of an ecosystem. Thus my corporate friends could easily convince themselves that the monarch butterfly paper could not be nearly so damaging to the science and industry interests of plant biotechnology as the tall tale of toxic potatoes. But as it turned out, we were all wrong in our wishful thinking. The *Nature* paper and subsequent press coverage provided a strongly focused rallying point for Greenpeace and other environmental organizations. They used the opportunity for colorful protests in which some of their more dramatic members would dress up as butterflies, flutter around a guy dressed in an ear-of-corn suit that had a giant X on it, and then keel over. It was fodder for the camera, and the account of dying butterflies hung in the air for several weeks. Furthermore, the public remembered the story for months as, for the first time, the ecological debate about GM plants came to North America with fervency. In 1999, the controversy had jumped the pond from Europe and Americans began to seriously question GM plant biosafety. While the controversy hit hard in 1999, it would take a succession of scientific papers over the next two-and-a-half years to learn the real truth about the monarch butterfly.

The Losey et al. Paper

After my experience with the gloom-and-doom lunch, I have to admit that I was quite surprised by the brevity of the *Nature* paper.[6] It described a single unreplicable experiment and brief results: Bt corn pollen force-fed to monarch butterfly (*Danaus plexippus*) larvae caused mortality. There was ultimately much more to the story, but the single experiment was all the science that would be published on Bt corn and monarch butterfly during the next year. The original paper was the scientific equivalent of shouting "fire" in a theater, and so the fire hoses were turned upon biotechnology. Equivalent results could have been obtained by spraying monarchs with the same Bt bacteria that organic farmers frequently use against insect pests instead of Bt transgenic plants, and conceivably, organic farmers would have been the monarch murderers instead of the biotechnologists.

Monarch butterflies are fascinating insects. They are large, beautiful butterflies that are the embodiment of everything stunning in nature (figure 6.4). Each year they make a round-trip journey from Mexico, where they overwinter, to the United States and Canada, where they feed and reproduce (figure 6.5). The life history of the monarch butterfly might be the only entomology some schoolchildren are ever taught. The public's

Figure 6.4. Monarch butterfly adult. (Photo by Photos.com.)

Figure 6.5. The fall migration pattern of the monarch butterfly that occurs between September and November each year to overwintering sites in Mexico.

thoughts and feelings about the monarch butterfly are part nature study, part aesthetic admiration, and part heartwarming sympathy. Killing monarchs is nearly the emotional equivalent of hunting Bambi and clubbing baby seals. Never mind that the monarch butterfly is more closely related to my childhood nightmare insect, the corn earworm, and cockroaches than it is to deer and seals. Human feelings and perceptions often part ways with scientific knowledge, and the outrage that was poured out toward biotechnologists as a result of the publicity of Losey's paper was partly because people felt sorry for the monarch butterfly.

In their temperate summer habitat, monarch larvae are hungry for their sole food: milkweed leaves. Common milkweed (*Asclepias syriaca*) and its relatives grow in and around cornfields. Just like all other caterpillars, monarch larvae live to eat. The larvae's goal is to put on weight so they can successfully pupate and morph into adults, the lovely winged creatures that are the posterchildren for insects everywhere.

Bt corn pollen, which does contain small amounts of Bt toxin, could conceivably drift to, and land on, milkweed leaves, where the monarch butterfly larvae could potentially eat their tainted food and die. Conceivably, this is the reason Losey performed his experiment.

In a lab experiment, Losey and his colleagues simply dusted milkweed leaves with an undisclosed amount of Bt-containing corn pollen and watched what happened to the caterpillars (figure 6.6). They observed decreased feeding, growth, and survival rates in exposed larvae compared to larvae that consumed leaves dusted with non-transgenic corn pollen. From this single lab experiment, the authors extrapolated to the 100,000 square miles of corn, milkweed, and monarch habitat in the midwestern United States to conclude that Bt corn poses a danger

Figure 6.6. Monarch butterfly caterpillar and high density of corn pollen on milkweed leaves representative of conditions imposed by Losey's research group in their 1999 *Nature* article. (Photo by Kent Loeffler.)

to monarch butterfly populations. To quote Losey from the Cornell University press release that publicized the *Nature* paper: "We need to look at the big picture here. Pollen from Bt-corn could represent a serious risk to populations of monarchs and other butterflies, but we can't predict how serious the risk is until we have a lot more data."[7] That statement sent mixed signals about the scope of possible harm. But, clearly, many lay people and activists assumed the worst. The same could have been said for Bt sprays, though, since the same toxic effect would have been observed.

Because of the narrow scope of the study, many scientists in this research area questioned the validity of the experiment and paper itself, even with Losey's caveat that they needed more data. There were criticisms that the methods were not reproducible and that the "non-choice" feeding experiments did not represent, in any way, the conditions found in real cornfields. Furthermore, the amount of pollen deposited was obviously artificially high (but unknown). So the door was wide open for follow-up studies, which were done by a host of other researchers.

Second Monarch Paper

More than a year passed before the second Bt corn/monarch butterfly study was published.[8] Entomology professor John Obrycki and graduate student Laura Hansen Jesse at Iowa State University performed follow-up experiments with the same monarch/Bt corn pollen/milkweed experimental system. Although they were careful to avoid some of the pitfalls of the Cornell experiment, and the experiments included more field-based approaches, these too, were not really field studies. Seeing another, albeit better, set of laboratory-based experiments was disappointing to everyone interested in truly understanding if and how Bt corn was impacting monarch butterfly biology. There was a lot at stake, and there was a great deal of anticipation that after the 2000 field season an honest-to-goodness field study would be forthcoming. It turned out that we'd have to wait an additional year for that to happen.

Jesse and Obrycki performed bioassays that were laboratory-based, where the larvae were not given food choice—a Losey et al. redux. However, pollen amounts on milkweed plants were quantified, which was a notable improvement from the Cornell study. In addition, the sites that were chosen for the work were located in Iowa, in heart of the U.S. corn belt. The researchers measured corn pollen loads on milkweed plants in and around cornfields. So while there were no real field experiments per se, the group did perform some initial needed field surveys.

The Jesse and the Obrycki study tested different transgenic events on monarch butterfly neonate larvae and found both acute and sublethal effects.[8] The transgenic lines used were the 176 and Bt11 events in two different corn hybrids each. This choice of event is important because no one knew which event—specific transgenic line—the Cornell researchers used (we now know it was Bt11). The Iowa State study also used a proper experimental design. Good scientific practice in plant biotechnology demands that the transgenic events used in experiments be identified, so studies can be replicated and properly interpreted. And in this case, the two transgenic corn events in question used dissimilar promoters with distinct activities. The 176 event was known to produce more than 100 times more Bt toxin protein in pollen grains than Bt11. Most researchers in the field assumed that Losey et al. used event 176 because it possessed unduly high levels of toxicity to caterpillars. Many researchers in field cried foul on that front because 176 was planted on very low acreages, which subsequently shrunk to zero 2 years after the original paper. However, most scientists in this area now believe that in both the 1999 and 2000 studies that the pollen grains were tainted with tassels (anther material), which contained significantly more Bt toxin protein than did pollen grains. Tassel contamination artificially elevated the apparent toxicities and negative effects of Bt corn on monarchs.

The most significant finding of the Iowa State study was the determination that very little corn pollen traveled any significant distance from the transgenic corn plants; almost none was detected 1–3 meters outside of cornfields. In addition, larval mortality seemed clearly to be linked to pollen load, decreasing as load decreased, therefore indicating that the risk to monarch larvae would rapidly diminish with distance from cornfields.

The Field Studies Arrive; The Fat Lady Sings

Finally, in 2001, 2 years after the original paper, a barrage of papers on Bt corn and the monarch butterfly were published together in the prestigious journal *Proceedings of the National Academy of Sciences USA*, also known as *PNAS*.[9–14]

The publication of the 1999 *Nature* paper triggered researchers to collaborate on research about Bt corn and butterflies. Rick Hellmich, a scientist with the USDA Agricultural Research Service at Iowa State

University, led an impressive collaborative effort. Also, a group of scientists produced a road map for how to determine the real impacts of Bt corn on monarch butterfly populations. They established a broad steering committee that consisted of two USDA administrators, a person from industry, a person from a research university, and a representative from the Union of Concerned Scientists; this committee evaluated research proposals that were funded by the Agricultural Biotechnology Stewardship Technical Committee (ABSTC). The ABSTC requested and received money from industry that was pooled and distributed as directed by the monarch steering committee. Most of the working group of scientists received some support from the ABSTC. Hellmich's group did not receive any of this funding for research, and, likewise, several others did not request any funds. These scientists used other funding besides those from the ABSTC to do the research.

While nearly all researchers have collaborators, the degree and scope of mission-oriented collaboration on this project was especially expansive. They agreed that the goal was to deliver a risk assessment that was based on the collective data of the group. It was important in the effort to replicate most experiments in three locations so as to include the natural variation that monarchs over their entire range would typically experience. The group wanted to collect even more data than it did, but it ran out of time and money.

The *PNAS* papers examined the toxicity of Bt toxins and Bt corn pollen and tassels on monarch butterflies.[9] They noted patterns of corn pollen deposited on milkweed in and near cornfields and the overlap of monarch caterpillars and corn pollen in fields and during the growing season.[10,11] Field studies showing the response of caterpillars to realistic levels of Bt corn pollen were finally performed,[12] and a grand synthesis of analysis of the studies in the form of a risk assessment of what it all means was published.[13] Finally, additional work was done on the now defunct 176 event, indicating that it would indeed have had non-target effects on butterflies.[14]

These papers authoritatively put the monarch butterfly biosafety question to rest—at least until new insecticidal corn varieties that control lepidopterans are produced. The funny thing was, the popular press had very little to say about this highly anticipated and thorough effort.

The studies published in *PNAS* first attempted to replicate the lab-based experiments published in the two papers in 1999 and 2000. In addition, the research team tested all the Bt corn events that were on the market, plus other experimental events as well. In keeping with the other studies, no-choice assays were performed, so the insects were forced to eat Bt toxin, which determined the amount of a lethal dose. Hellmich et al. noted that if tassel parts were included in the assay along with pollen, the toxic effects were increased.[9] For example, if there were 1000 pollen grains/cm^2 of milkweed leaf area (and that is a lot of pollen) using the Bt11 event, 97% of the monarch larvae survived for 4 days, but only 17% survived if the pollen grain samples were contaminated with tassels. The Bt11 event (one of the most widely planted types of corn) tassels produced 100 times more Bt toxin than did pollen. Yet in the field, tassels don't drift onto milkweed nearly as much as pollen does. Furthermore, when monarch larvae ate high densities of corn pollen (Bt or non-Bt), this led to a decrease in weight gain. But it is clear that insects can eat around large masses of pollen on milkweed leaves, similarly to how we can eat around wormholes in apples. The only transgenic corn event that consistently killed larvae at low densities was 176, which was never grown very much by farmers in the first place. So, at first blush, on the basis of lab-based studies only, when monarch caterpillars were forced to eat milkweed with relatively low amounts of Bt corn pollen deposited (without the tassels), there seemed to be little risk of harm. But to answer the questions of how much pollen and tassels are actually deposited on milkweed leaves, real-life field experiments had to be performed. Even if proper lab experiments had been performed in 1999, though, the alarm would not have been nearly so great.

Corn pollen is quite heavy and doesn't travel far. Pleasants et al. examined the spatial patterns of corn pollen deposition in time and space at various locations in the United States and Canada.[10] The highest average pollen grain deposition of corn pollen on milkweed existed within confines of cornfields and was only 170.6 grains/cm^2. Two meters outside the cornfield, the average deposition decreased to 14.2 grains/cm^2. Within all the cornfields surveyed, only 5% of the samples had counts of greater than 600 grains/cm^2. In only one sample was there a toxic concentration of pollen: 1400 grains/cm^2. So, in and around real cornfields, monarchs seldom experience acute exposure to Bt corn pollen. In the case of 1400 grains/cm^2, there had been little rain during the period. Rainfall is known to wash pollen grains off milkweed leaves. For this reason, the pollen distribution is never uniform on all leaves. Milkweed leaves located on upper portions of the plant had up to 50% less corn pollen compared with middle leaves. Pollen densities on lower leaves were between 50%

and 100% of that on middle leaves. So, if you were a monarch butterfly caterpillar, which leaves would you prefer to eat? What if Bt corn pollen or any corn pollen is on a particular milkweed plant? And can you tell the difference? It turns out that female monarchs most often lay their eggs on upper leaves, which is where monarch caterpillar neonates hatch and feed first. So caterpillars are exposed to less than average amounts of Bt corn pollen on milkweed plants by eating the upper leaves of the plant where they hatch.

Oberhauser et al. also surveyed many corn/milkweed fields throughout the United States and Canada to determine whether there was much overlap in time between anthesis (when pollen is shed) and when monarch larvae feed.[11] In Iowa, there are many more monarchs in agricultural fields compared with more northern locations. And in northern locations there is more overlap of monarchs and corn pollen compared with more southern locations. So, while there is no doubt that monarch caterpillars have an opportunity to eat Bt corn pollen wherever it is grown, the time of anthesis in any particular field is only 1–2 weeks. Monarch larvae are in the field before summer starts and after it ends. Few individual insects would ever have an opportunity to be affected, even if there were a high dose of Bt toxin delivered onto milkweed leaves, which there is not. It is clear that Bt corn is not a limiting factor controlling the size of monarch populations in the field.

Finally, Stanley-Horn et al.[12] examined the effect of Bt corn pollen on monarch caterpillars in the field in several fields in the United States and Canada.[12] They compared the effect of Bt with the effect of a sprayed insecticide, λ-cyhalothrin, which is commonly used to control economically important insects on corn. The only Bt corn event that had any negative effects (in this case, less caterpillar weight gain) in the field was event 176. Bt11 and Mon810 pollen, when at concentrations of 97 grains/cm^2 on milkweed leaves, had no effect on monarch survival, when compared with non-Bt pollen. In one instance they found a Bt11 site as high as 586 grains/cm^2 with no accompanying ill effect on monarchs. Their studies integrated mixing tassels with pollen in the field. They did not find any tassels on milkweed leaves. However, in non-Bt fields treated with λ-cyhalothrin, there were significantly fewer monarch butterflies, demonstrating that monarchs would fare better in Bt corn fields compared with those insects in conventional corn fields sprayed with chemical insecticides.

In their risk assessment paper, Sears et al. concluded that Bt corn caused little risk to monarch butterfly populations.[13] Further analysis of the Bt corn/milkweed/monarch butterfly system shows one other important feature. Farmers view milkweed as a mere weed in cornfields and not as monarch food, and they would rather milkweed not be present to compete with the corn crop. Farmers have historically removed milkweed and other weeds as they are able. Iowa State University weed scientists have demonstrated that less than half of surveyed plots in Iowa cornfields contained milkweed.[15] The milkweed plants that were present were generally growing at low densities. So farmers had done a good job in removing these weeds. In contrast to cornfields, milkweed is found much more frequently on roadsides in Iowa, where we would presume there is no specific removal. More than 70% of plots surveyed contained milkweed plants, and the area occupied by milkweed along roadside samples was more than three times larger than areas within the cornfields. So, there is a lot more milkweed by Iowa roadsides than in cornfields. A twofold problem therefore presents itself that has nothing to do with biotechnology: herbicides and cars. Weed killers and automobiles kill more monarch butterflies than does Bt corn pollen. Farmers destroy monarch food with herbicides, so that monarch butterflies are forced to roadsides to find food. And we've all had the misfortunate of seeing flying monarch butterflies splatting on the windshield. The *PNAS* papers indeed looked at the big picture of an apparent risk of biotechnology in the practice of field-level agriculture.

Big Picture

The big picture shows there is a significant environmental upside of planting transgenic Bt crops—especially cotton. The obvious environmental advantage is that Bt crops allow fewer chemical insecticide applications, which indiscriminately kill monarchs and other beneficial insects. In field corn, there has been no apparent reduction in insecticide usage since Bt varieties were introduced. However, Bt corn has enabled farmers to garner greater yields and cultivate less land to produce the same amount of crop. Less land under the plow is a tangible environmental benefit translating to less soil erosion, and this benefit is accompanied by the use of fewer fossil fuels and collateral air pollution from working the land. Higher yields means less land is required to be planted in crops. In turn, this results in greater biodiversity on fallow land and land that is allowed to naturalize compared with land that is planted in crops.

The monarch butterfly issue has been nearly laid to rest. While it is true that Losey and a few colleagues are still worried about monarch populations being vulnerable to Bt corn pollen (at least in 2002),[16] most scientists are convinced that the data from the battery of *PNAS* papers demonstrate there are negligible risks in the system. To leave a back door open, Losey and colleagues state, "even if mortality caused by Bt corn is only a small fraction of the total mortality caused by other sources, it could theoretically be enough to send monarch populations into a downward spiral."[16(155)] But this statement seems irrelevant when conventional insect control strategies (including chemical insecticide applications) that have been used for years are taken into account. Wouldn't this downward spiral caused by insecticide sprays have occurred long before Bt corn was grown? After all, it has been shown that chemical insecticide applications are more toxic to monarch butterflies than Bt corn in the field. But old images don't die quickly. Two years after the flurry of *PNAS* papers dismissed the risk of Bt corn on monarch butterflies, protesters were once again donning their corn and butterfly suits and reenacting a fallacious drama at a prominent biotechnology conference.

In conclusion, while the monarch butterfly episode caused a significant downturn in agricultural biotechnology, one small consolation prize was the biosafety reality check that visited industry. The companies realized that they needed to fund more studies on the ecology of transgenic plants. At present, industry still pays very little of the bill of independent biosafety research, but it is, at least, taking ecological and environmental effects more seriously now.[17] Finally, the butterfly episode also worked a little humility into the agricultural industry, which before this incident attempted to seemingly steamroll those opposed to its biotechnological revolution.

While the monarch butterfly story has been the most visible on the landscape of farm biotechnology, there are other potential side effects that have caused concern about insect-resistant GM crops. Unlike the monarch, which is an insect that is prettier than it is ecologically important, the other insects that could be potentially harmed by Bt crops are known to have beneficial properties on the farm. It is certainly undesirable to decrease the populations of beneficial insects that play a role in controlling pests. What are the other nontarget effects of Bt plants that could erupt into the next perceived disaster? Do we know enough to dismiss the next "monarch candidate" as implausible? Exactly how much do we know about field-level side effects that Bt crops have on beneficial insects?

7

Better Living through Biology

Not Killing the Good Insects by Accident

It should not be surprising that after weeds, insects are the most damaging pests in farming. Humans have faced monumental challenges against insects through the millennia, both inside and outside of agriculture. Locust infestations of biblical proportions have been the pinnacle of mankind's war against insects—typically ending with humans on the losing side. Insect-vectored diseases such as the flea-transmitted bubonic plague crushed Europe in the 1300s. Episodes of malaria carried by mosquitoes inspired the use of DDT and other chemical insecticides. These types of early successes of chemical insecticides led to their wide adoption on farms in the 1940s. Around the same time, there was redoubled effort to breed plants for increased resistance against insects. In the decades that followed, despite well-documented problems with their overuse, chemical insecticides significantly increased crop yields, even more than plant-based resistance had achieved through conventional breeding. Toward the end of the twentieth century it became clear that transgenic insect-resistant plants had the potential to combine the best feature of chemicals (in this case, protein-based insecticides produced in GM plants) with plant resistance. At the end of the twentieth century, it was finally possible for crops to defend themselves against insect pests. As the twenty-first century began, transgenic plants had a clear and integral role to play in pest management.

Integrated Pest Management

The current paradigm of insect control in farming is integrated pest management (IPM). IPM strives to manage weeds, insects, and disease-causing pests using many available synergisms. The toolbox includes chemicals; plant resistance; beneficial insects, such as natural insect predators that eat insects that damage crops; and cultural practices, such as field selection and crop rotation. The goal of IPM is to use all prudent inputs and aspects of agricultural ecosystems to prevent the buildup of economically damaging levels of pests. The goal of IPM does not include totally eradicating insects and weeds, but minimizing economic losses caused by pests. IPM has many interlocking pieces that theoretically work in the best economic interest of the farmer while creating a small environmental footprint. IPM was proposed in the late 1950s as an alternative to the emerging panacea of chemical insecticides. The practice of IPM tends to value chemical pesticide application as a lower-tiered means of preventing pest populations from spiking. One part of IPM might be yearly crop rotations—for instance, between cotton and soybeans, so that the insects that eat cotton plants are faced with a field of soybeans every other year. Another IPM practice is the establishment of beneficial insects, such as lady beetles, that will consume aphids and other pest insects. While these parts work together, chemicals are still sometimes needed to provide effective control. For instance, cotton growers in the Mississippi Delta rely on multiple applications of chemical insecticides to make a crop (figure 7.1). For that reason transgenic crops that express insect resistance genes are a new piece of the IPM puzzle that is welcomed by growers whose previous top gun in the battle against insects was the application of chemical insecticides.

The idea that transgenic plants can at least partially substitute for synthetic chemical insecticides is quite attractive on many fronts. One of the leading ecological reasons is that transgenic plants should not have nearly as many nontarget effects as chemical insecticides. Chemical insecticides are commonly broad spectrum in action; they do not differentiate between good and bad insects. Insecticide sprays are like rain—everything gets wet. Even though GM plants are much more precise with regard to killing a small range of insects, they have also been suspected of having negative nontarget effects. Some detractors of biotechnology claim, however, that the monarch butterfly situation was only the tip of the iceberg of nontarget effects of GM plants and that insect-resistant transgenic plants could have a much larger ecological footprint than initially presumed. These potential negative ecological side effects include killing beneficial insects and harming the microscopic soil flora and fauna that largely compose the unseen underground ecosystem within agricultural fields and beyond.

Figure 7.1. Photo of a crop duster spraying insecticide on a cotton field. (Photo by Photos.com.)

This concern is important to investigate because IPM is a practice agriculturalists desire to keep intact. Chemical insecticides accidentally kill beneficial insects as well as pests. GM plants should only be adopted if they improve IPM. If GM plants were to jeopardize IPM, perhaps they should be reexamined as an agriculturally useful technology.

Killing the Good Insects by Accident

We closely examined scientific methods used to determine whether Bt corn pollen is lethal to a very high-profile insect, the monarch butterfly. Likewise, throughout this chapter it is important to keep in mind the ways lab-based experiments are not particularly predictive of real life in agricultural fields. This is not to say that lab-based ecological and toxicological experiments are without value. We must simply remember that they can be easily misinterpreted and their results extrapolated beyond their reasonable scope. Not that the scientists who do the work are the ones who do the misinterpreting; after scientific findings are published in a journal, they are often considered fair game, and bully pulpits abound. So now imagine studying systems of insect interactions that are considerably more complicated than the Bt corn pollen, milkweed, and monarch caterpillar system—not just one type of crop, weed (or insect food), and insect. Instead of a linear vertical relationship in the monarch system, the components can often be weblike and empirically and logistically difficult to study. To understand certain ecological interrelationships among organisms, several researchers have studied experimental systems involving tritrophic interactions.

"Trophic" refers to nutrition, so "tritrophy" means there are three levels of eating. Typically, at the lowest trophic level would be the crop plant—the base of the food chain, and, in this case, a GM plant. The next trophic level would be the pest insect. In the absence of a transgenic plant, chances are it would be controlled by IPM, including chemical insecticides if necessary. The third trophic level might be occupied by a beneficial insect that eats the pest insect. We will examine two experimental tritrophic systems that were used to test for the side effects of GM plants.

Canola, Worms, and Wasps

Tanja Schuler and colleagues used Bt canola plants produced in my lab to perform insect behavior experiments using choice feeding assays.[1] Unlike the no-choice assays used in the early monarch studies, choice experiments allow the pest insects, in this case, *Plutella xylostella*, the diamondback moth, to choose their food source (canola) from a number of transgenic and nontransgenic selections, a kind of a limited buffet for insects. Diamondback moth caterpillars prefer to eat the leaves of all kinds of mustard plants, including canola. At the third trophic level was the parasitic wasp (*Cotesia plutellae*) that makes its living off diamondback moth larvae. While we might imagine all wasps as the large, menacing, stinging variety, *Cotesia* are tiny insects dangerous only to diamondback moth caterpillars. They lay eggs on their victim, and their offspring eat the caterpillar. *Cotesia* are considered to be beneficial insects by canola farmers. The researchers investigated whether the beneficial insects were harmed by Bt canola. In their study, they showed that the parasitic wasp prefers happy diamondback moths. What exactly defines happy diamondback moths, and where might we find them? *Cotesia* wasps have something akin to a radar system that helps them sense damaged canola leaves. They smell volatile odors emitted from chewed-up canola leaves, so where damaged canola leaves are found, there is a good chance that feasting diamondback moth caterpillars will be present. This research group did something very clever to test whether Bt canola was harmful to the beneficial *Cotesia*. They used two kinds of plants: non-GM canola, and Bt canola in their lab-based choice experiments. They also used two different strains of diamondback moths: Bt-susceptible diamondback moths, which die after ingesting Bt protein, and a Bt-resistant strain of diamondback moth. Bt-resistant moth larvae can eat Bt canola with impunity because the Bt toxin is inert to them. Bt-susceptible diamondback moth larvae are only happy (and alive) when eating non-Bt canola. Bt-resistant caterpillars are happy and healthy eating either kind of canola. The inclusion of all these experimental factors helped tease apart the true effects of the Bt plants.

The experiments demonstrated that it did not matter if the plants were GM or not, as long as the diamondback moth could ingest leaves without toxic effects.[1] If diamondback moths were able to eat the canola, the wasps would come to attack and kill them. The inclusion of Bt-tolerant larvae in the experiment uncovered that the key factor of *Cotesia* predation was actually plant damage. The parasitic wasp experienced no ill effect from eating Bt protein via the diamondback moth caterpillars, but there was no reason to expect that the wasps would be harmed, since they are not closely related to butterflies. So if Bt-resistant caterpillars were to arise in the real world where Bt canola was planted, *Cotesia* wasps would help control these caterpillars as well. The Bt toxin in the transgenic canola did not harm the beneficial wasp in any way.

Corn, Corn Borers, and Lacewings

A second series of tritrophic interaction experiments studied in the lab involved Bt corn as the plant and primary food. The Bt gene in corn protects corn stalks from the pest insect, the European corn borer (*Ostrinia nubilalis*), which normally consumes the inside of the corn stalk and causes plants to collapse. The predator at the third trophic level is the green lacewing (*Chrysoperla carnea*), which looks extraordinarily vicious at the larval stage but metamorphoses into a very pretty adult (figure 7.2). In addition to eating European corn borer larvae, green lacewings also consume other caterpillars and aphids. Angelika Hilbeck and colleagues in Switzerland have been studying this Bt corn/pest insect/lacewing system for several years and have provided valuable information about the biology

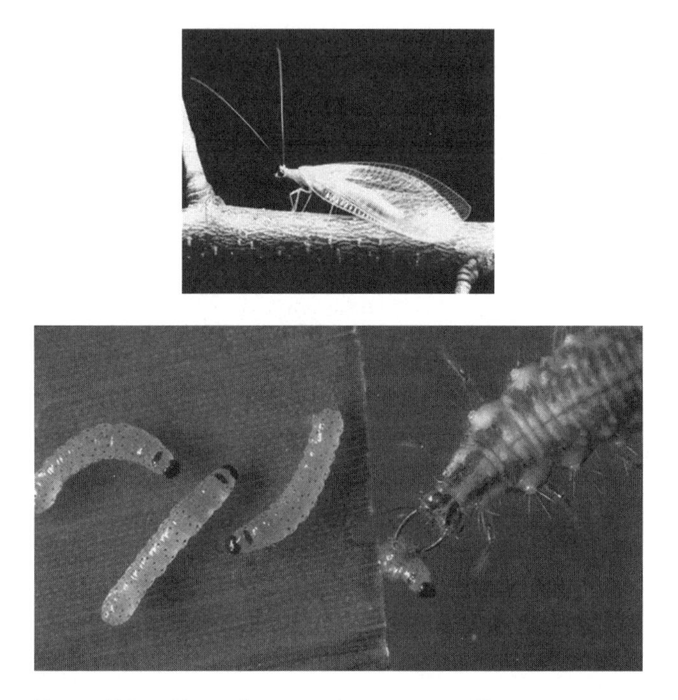

Figure 7.2. Green lacewing larvae eating European corn borers on the backdrop of a corn leaf (top), and the adult (bottom). (Photos by Matthias Meier.)

of the system.[2–4] Their goal has been to determine whether green lacewings would be harmed if they consumed corn borers that had ingested Bt transgenic corn. As a control, they also directly fed lacewings doses of Bt protein incorporated into artificial diet. They found that the lacewing larvae suffered higher mortality rates from feeding on European corn borers or other nontarget caterpillars reared on Bt corn compared with those fed the non-Bt corn. They also found that when high levels of Bt protein (about 20 times the amount found in Bt corn) were incorporated into the artificial diet, there were higher mortality effects to the corn borers. One interesting and curious result was that the highest level of lacewing mortality occurred when lacewings fed upon corn borers that ate Bt corn, and not in the treatment that had Bt insect chow, an unexpected effect if Bt protein was the primary cause of death. One likely explanation is that the Bt protein made the corn borer caterpillars sick and therefore less nutritious to lacewings. An analogous question is, would you rather eat a healthy chicken or one that died with pneumonia or one that died of food poisoning? It seems that lacewing larvae fared better consuming healthy corn borers.

Nonetheless, there is apparently a specific Bt Cry toxin effect on green lacewings, a beneficial insect species. Hilbeck believes we need to know more about the sublethal effects of Bt toxic proteins on secondary and tertiary trophic levels and their ecological effects, one motivation for continued study in this system.[2–4] But as discussed in chapter 6, we've seen that significant lab-based results do not always relate to the real world of agricultural. Although it is certainly important to understand the effects transgenic plant-produced Bt proteins have on beneficial insects in the lab, it is even more important to understand how populations of beneficial insects are impacted by transgenic plants, insecticidal sprays, or even Bt bacteria that organic farmers use to control damaging insects. We know little about how lacewing populations fare in any of these insect control methods.

The ultimate way to understand the side effects of any pest control in the real world is to compare it to all relevant methods in the same field locations. These comparative field experiments still need to be performed; however, few ecologists have performed these kinds of experiments. There are several reasons for this dearth of research. The first is expense. There is little public funding for GM plant ecological research, especially in the United States. It is much less expensive to perform lab experiments. The second is that interdisciplinary teams are needed, which takes more effort and administrative skill than is always available, especially in light of the expense. Third, the experiments are necessarily large and complicated. Later I examine one of the few field experiments that have been performed of field-level ecosystem effects of Bt corn on nontarget insects. But first, we will take a look underground.

Bt Protein: Out of the Plant and Into the Ground

Microbiologist Guenther Stotzky and his colleagues at New York University observed that Bt toxins would bind and persist in soil, especially in clays.[5] This finding suggests potential risks to nontarget organisms in soils where Bt crops are grown. Indeed, Bt toxin proteins bound to soil were found to be toxic to susceptible caterpillars when they were fed the soil.[6] And the Bt protein did not dissipate once it bound to clay soils. Stotzky and colleagues found that insecticidal activity of Bt can be maintained at least for 234 days.[7] They also found that root exudates from Bt corn contained Bt protein when grown in hydroponic solutions as well as when grown in soil.[8] It seems that Bt corn could dump Bt protein from its roots into soil, where the toxin could persist for perhaps up to a year or more, retaining its insecticidal activity; in other words, it could kill susceptible insects if they ingested it in the soil. This scenario was especially plausible in soils composed of kaolinite, a type of clay. Although the long-term persistence of Bt toxin in the soil is notable, this group, so far, has not unearthed any negative or unexpected effects from Bt's persistence in soil. For example, Bt protein in soil does not harm earthworms (even though it is found in their guts), and it is not taken up by non-GM plants when they are grown in soil where Bt corn was previously grown.[9] In addition, other Bt-transgenic plants do not secrete Bt proteins into the soil. Stotzky and his research team have been leaders in examining the exudation of Bt proteins into the rhizosphere and their binding onto clayey soil. But these results contrast with those from several other studies that have shown rapid degradation of plant-expressed Bt proteins comparable to the rate of degradation of Bt proteins in microbial products.[10,11] There are still some aspects of this experimental system that need to be better understood.

The many questions raised by these various studies demonstrate a clear need to analyze possible nontarget effects caused by GM crops. For such research to be entirely relevant to farmers and ecosystems, it needs to be framed in the context of current practices in agricultural systems. To reach that goal, field experiments are necessary.

There are two possible outcomes from lab-based research of the type we've examined here. Experiments can lend credence to potential risks associated with GM plants or, alternatively, to risks being absent or negligible. These are, of course, always predicated on good experimental design. If a result is negative, indicating low risk, there may be no need for additional field research, such as in the case of the *Cotesia*/diamondback moth example. In contrast, positive results (possible or probable risks) indicate that fieldwork should be performed to test whether the effect is real or an artifact of artificial conditions. It is a long way from the lab to the field. One example that illustrates that distance can be found in the results showing that Bt toxins can survive intact in water.[12] Stotzky's group found that Bt protein could be collected in water leached through Bt-contaminated soil. Once again, Bt protein was active on target insects. There was a diminished toxic effect when Bt corn (rather than pure Bt

Cry protein) was used to test for leaching. The potential side effects of Bt toxin movement into water could be alarming; it is conceivable that Bt toxins could be traversing through well water, rivers, and streams killing nontarget caterpillars as they ingest the toxin in water. The researchers did not examine the effects of Bt sprays that might be used by organic farmers, another appropriate treatment. But, as Stotzky and colleagues point out, there are several field-level factors that would determine plausibility of these side effects, such as rain, snowmelt, cultivation practices, soil type, and crop management.[12] In addition, the soil microbes in agricultural fields may affect the binding of Bt proteins to soil and leaching into water. Microbes play a role in the degradation of the protein itself. It can be difficult to read the complex road map between the lab to the field.

Many studies have tended to focus entirely on recombinant proteins that would be synthesized in transgenic plants. It is obvious why one would research Bt proteins, but the results found so far make me curious about ordinary plant proteins that are naturally secreted into soil. What are their effects on biological systems in and out of agriculture? It is not obvious to me why we should solely be interested in Bt endotoxin proteins, which have very narrow host ranges, when there are many more toxic proteins that plants naturally produce. And what about digestive enzymes that we use to unclog drains and proteins in food that find their way to sewers, including those from a diversity of organisms, from sewer rats to corn? What are their persistence patterns and side effects? The example of Bt protein persistence in soil is a good example of why relevant field experiments are necessary as a follow-up to lab studies as quickly as possible. If there are effects, we should certainly know as much as we can about them, and if there are none, then we should be aware of that fact as well. It is important not to jump to hasty conclusions on the basis of lab experiments only.

Out of the Lab into the Real World of Agriculture

As we've seen above, assessing real-world side effects of transgenic plants is a difficult process. Populations and communities of tiny insects, spiders, fungi, and bacteria exist in agricultural fields and functionally require significant space and interactions to exist. Saxena and Stotzky have shown that Bt from corn root exudates has no effect on earthworms and microbes such as protozoa, bacteria, and fungi in soil.[9] Some microbial species are relatively easy to study in the lab if they are culturable. It is much more difficult (or impossible) to assess ecological structure of communities of organisms in the lab. Although we've talked generally about moving from the lab to the field, there are some practical and logistical problems in doing so. We've already discussed financial hurdles and the interdisciplinary teaming needed to do the best experiments—these hurdles can be nearly insurmountable for the university-based researcher. It is difficult to assemble the multidisciplinary team, raise the funds for such a study, and then be patient enough to wait years for the few publications that will come from the study. Professors need to publish more than one paper every 5 years. Because of these significant hurdles, I only know of one comprehensive study that has been published in a peer-reviewed journal to help us understand the side effects of growing an insecticidal transgenic crop on the farm, and industry scientists did it. Other than this one, there are several papers that have not yet been peer-reviewed and published in the scientific literature. In addition, the largest series of field experiments have been performed in the United Kingdom, which has had the foresight to heavily fund farm-scale transgenic plant evaluations. These, however, focus on herbicide tolerance, not on Bt, and will be briefly discussed later in this chapter. But, as far as I know, they are the only ones of their type to be initiated by governments and academicians. I think that we in America could learn a few things about funding science in this field, especially given the fact that most of the commercial GM plants are born in the United States, where field testing has been easier from a research and regulatory standpoint than in most other parts of the world.

French Cornfields

The most inclusive set of information we have on the field-level side effects of Bt plants comes from transgenic corn experiments in France performed by Syngenta scientists (when it was part of Novartis).[13] The Burgundy region of France is home not only to grape vineyards, but also to cornfields. The cornfield studied was 15.1 hectares (a little over 37 acres) and was subdivided into 10 plots that varied from just under 3 acres to just over 4 acres. These plots differed with regard to the type of corn grown (Bt or non-Bt) and the kind of chemical treatment. The types and numbers of plots were as follows: (1) three plots contained Bt corn (Bt Cry1Ab, which controls European corn borer caterpillars) with no insecticide applications; (2) three control plots of non-Bt corn (same variety as GM corn, but no Bt) also not sprayed with insecticides; (3) two plots of non-Bt corn sprayed with a Bt insecticide formulation called Delfin, a commercially available Bt spray that organic farmers would use to control a number of different insects on corn (it contains several different Bt Cry1 and Cry2 toxic proteins); and (4) two plots of non-Bt corn sprayed with a synthetic insecticide (a pyrethroid called KarateXpress). Both the Bt and synthetic insecticides were sprayed once in mid-July when the plants were still fairly small. The choice of insecticides and application timing was realistic for commercial corn production. Karate is often used to control corn borers, but a Bt spray is not typically used because it is not as effective against the target insect compared with a synthetic insecticide such as Karate. However, the Bt spray was a realistic experimental control for the Bt transgenic plants.

I remember hearing a presentation from one of the scientists involved in this study at a biosafety meeting in Europe. Because it was the first paper of this type, the speaker had an enthralled audience which on one hand was extremely interested in the results, and on the other hand was seemingly anxious to critically examine the methodologies. As I recall, many of the scientists in attendance would have liked to have seen a balanced experimental design in which each of the four treatment types contained the same number of plots (in this case, all having three instead of some having two). A balanced design would have given higher statistical power for significance testing. Some of the scientists also took issue with the insect collection methods. I say these things now to indicate that the Syngenta study, while not perfectly designed, had, by nature, many constraints, including landform, size, and logistical barriers intrinsic to doing field research. These constraints must be addressed in good field studies, and in this case they were.

Arthropods of all types were collected from the soil, plants, and air. Arthropods included insects such as flies, beetles, butterflies, bugs, and noninsects including collembola, daddy longlegs, spiders, centipedes, and millipedes. In total, 76 different taxa (types) of arthropods were sampled from the soil, 45 taxa from plants, and 71 taxa from the air. The objective of the experiment was to determine if Bt corn negatively affected any of these taxa and to answer the question of the decade: does Bt corn decrease biodiversity among arthropods on the farm? It is important to note that among these arthropods were beneficial insects such as ladybugs, lacewings, predatory wasps, and pollinating insects such as bees.

These types of insects have been at the center of so many lab-based biosafety studies.

By far, the most interesting aspect of the paper is that it was filled with negative data, showing no harmful side effects from Bt corn. The Bt corn plots, when compared to the non-Bt corn control plots that received no insecticide application, did not differ in arthropod types or quantity. Bt corn had no side effects on nontarget species.

There were no nontarget effects of any Bt treatment, including Bt corn and Bt sprays when compared with the control among soil-dwelling arthropods, such as those that might be affected from Bt leaching out of the soil. The synthetic insecticide spray, Karate, did show effects on certain arthropods 14, 27, and 43 days after spraying. Millipedes, centipedes, a couple of kinds of spiders, and a type of beetle were harmed. It is interesting to note that no treatments, including the synthetic insecticide, resulted in a long-term decrease in biodiversity—that is, all the types of arthropods were not different in all the treatments.

A somewhat similar result was found with regard to plant-dwelling arthropods. Bt corn and Bt sprays had no negative effect on the biodiversity of the arthropods as compared with the control, while the Karate treatment resulted in relatively less arthropod biodiversity beginning only 2 days after treatment. The authors interpreted this difference as the result of increasing arthropod population sizes in the control plots around the first of July, just before spraying. In contrast with ground-dwelling arthropods, both the treatments of Bt sprays and Karate spray had negative effects on numbers of arthropods in certain groups. The two types of sprayed insecticide plots had fewer parasitic wasps, leafhoppers, ladybugs, soldier beetles, flies, and spiders compared with the control. The biodiversity and numbers of organisms rebounded after 15 days, demonstrating that the negative effects of spraying on biodiversity were temporary. While many individual insects were killed when sprayed, the survivors either reproduced and/or new individuals moved in from another field, replenishing the population. There were no biodiversity effects of any of the treatments on any nontarget flying arthropods.

These results will come as no surprise to those familiar with farming practices, insecticides, and GM plants. The effects (or the lack thereof, mostly) observed are completely expected when farming practices and the biology of plants and arthropods are thoroughly considered. Bt Cry toxic proteins have a very narrow range of toxicity; each toxin only kills a specific kind of insect. The Bt cry1Ab toxin only kills certain caterpillars, and only when corn tissue is eaten. There is no overdrift of Bt toxin produced by GM plants as there is when one sprays insecticides. Transgenic plants produce toxins that are more or less targeted to where the pest insect lives and eats, which, again, contrasts with sprayed insecticides. Sprays kill and stunt all kinds of insects whether or not they eat corn tissue. Chemical and Bt sprays are much harsher to "innocent bystander" arthropods than are Bt plants. Bt corn plots had fewer corn borers, as expected, but otherwise there were no effects on biodiversity compared with plots that were not sprayed and did not contain GM corn. It is interesting to note that while Bt sprays and chemical insecticides have been shown to kill nontarget arthropods, the negative effects on biodiversity of these treatments were temporary. At face value, the field-level data might cause the objective person to potentially question the long-range ecological safety of Bt sprays and chemical insecticides, but certainly not to question the safety of transgenic plants.

As Rachel Carson pointed out in *Silent Spring*, chemical pesticides have undesirable effects, and we should strive to move away from them by utilizing biology. Plant transformation with genes that code for proteins that have a narrow target range of insects seems to be the ideal compromise between good insect control for farmers and consumers of food (so we don't have worms, or even worse, part of a worm, in our food) and sustaining a decent environment. As the Candolfi et al. study demonstrates so vividly, even Bt sprays, which an organic farmer would find acceptable as a means for controlling insects, contain a cocktail of Bt toxins that broadens the scope of the types of insects that are killed.[13] So, while it is admirable that organic farmers want to use "natural" insect control measures, including Bt sprays, to tread lightly on the environment, it is clear that if they limit their toolkit to exclude GM crop technology, they will do more harm to the environment than the grower who relies heavily on GM crop technology for insect control.

Bt Cotton Fields

Another field-level study that addressed potential insecticidal side effects focused on Bt cotton.[14] In 2000, several industry and university scientists initiated long-term experiments in several cotton-growing states to compare Bt cotton expressing the Bt *cry1Ac* gene (Monsanto's Bollgard cotton) with conventionally grown cotton. Although interim results have only been reported in conference proceedings, the size of experiments and types of data collected are worth mentioning here because of their scope. Fields between 10 and 50 acres were paired, with the two types of plants and treatments. The first type contained Bt cotton with no insecticidal sprays, and the second contained non-Bt cotton sprayed with the kinds of chemical insecticides typically used on cotton. The researchers used two sampling techniques weekly to gather arthropods from vegetation. In most of the locations in 2000, most of the cotton, including conventional varieties, experienced only light herbivore pressure (conventional and Bt cotton) and did not require spraying of chemical insecticides. However, conventional fields in South Carolina did receive chemical insecticide treatments to control bollworms. The insecticide treatments decreased the numbers of natural enemies (beneficial arthropods). These beneficial organisms included predatory bugs, ants, and spiders. Head et al. also observed more aphids (harmful insects that suck plant juices) along with their predator, ladybird beetles (ladybugs) in the insecticide-sprayed fields.[14] The authors suggested that aphid populations increased as a result of decreased initial biological control in those fields (as the result of spraying), and that ladybugs migrated afterward to eat them. These results are consistent with data from the Candolfi et al. study.[13]

Concluding Thoughts on Corn, Credibility, and Chips

Let's turn our attention back to the Syngenta study since we have all the details laid out before us in a refereed paper. It is easy to perform a post hoc analysis to advise the Syngenta researchers about what they should have done differently in their experimental design and data collection. They admit in their paper that it would have been advantageous to collect more arthropod baseline data before the experiment began. Activists will certainly complain that it was "only" a one-year study and that plots of a few acres were used and that these are not large enough to be representative of large farms in the United States. Some people may also scorn a study that was performed by industry scientists only, as if to say their data were tainted or dishonest because of their industry support and employment. While I don't believe their science is compromised in the slightest, I would have liked to see Syngenta collaborate with people such as Drs. Schuler, Poppy, Hilbeck, and Stotzky on such a study in order to leverage their expertise. To Monsanto's credit, this is exactly what it is are doing in its multistate, several-year field studies. Monsanto tapped four university scientists at three different institutions as collaborators on its Bt cotton field experiments. As I have argued, if companies want to perform the best and most widely accepted biosafety research, they would be to their advantage to reach out to university scientists, mainly ecologists, who have expertise and perspectives that are largely lacking in companies.[15] Such collaborations not only would help the companies in their quest for sound biosafety data needed for commercialization (data they are required to present to regulatory agencies), but would also help the science to mature. Finally, such collaborations between industry and academia could shrink the philosophical chasm that exists between industry scientists and academic ecologists.

From my perspective, I can only see a win–win situation in greater collaboration and cooperation between industry and academia in agriculture and biotech biosafety research. To this end, perhaps industry should be taxed to create a pool of money that would be administered by a governmental agency, such as the USDA, to fund biosafety research. I have heard company scientists argue for their exclusive or near-exclusive stewardship of their products—that they ought to obtain most firsthand knowledge about the transgenic crops they produce, which is to say that companies should be performing the biosafety research. But that company line ignores their lack of public credibility when they are seen as policing themselves in the biosafety arena. Collaborating with government and university researchers not only increases the expertise and the scope of research for a company, it also might provide additional points of view when interpreting data for publication in scientific journals.

Implications

I think the Syngenta study on Bt corn will have significant implications for another type of Bt corn, one in which plant-produced transgenic insect control will have a much greater positive environmental effect: Bt sweet corn. The Bt corn that is currently grown on millions of acres in the United States and elsewhere is field corn that is used mainly for processed corn products for human food and animal feed, as well as for byproducts that are used in various industries. Sweet corn is the familiar, fresh, whole food we enjoy in the form of canned corn and corn on the cob. Field corn, relative to many other crops, is not heavily treated with chemical insecticides, and it has relatively few insect problems. The most important damaging insect in field corn is the European corn borer and then, to a secondary degree, corn rootworm and corn earworm. In contrast, a wide range of caterpillars can destroy sweet corn ears. Ponder the effect of the presence of insects and insect parts in field corn. If the corn stalk is bored out and the stalk collapses, it is still invisible to the consumer. Since the field corn kernels are ground up and processed, the consumer is never really aware that insect damage has occurred in infested ears of corn (nonetheless, I've often wondered about those little black specks in my lunchtime corn chips). But what happens if Aunt Sue or Uncle Steve finds a corn earworm in a can of corn, or just the head and half a body of a corn earworm? Chances are aunt and uncle will call the food company, grocery store, and personal injury lawyer, and give them an earful—and I'm not referring to ears of corn.

For quality assurance purposes, farmers must spray sweet corn fields multiple times—sometimes as many as 15 applications—per growing season with insecticides. Besides the potential for insecticide residuals on the food corn and compromised health of farm workers who apply pesticides, these sprays almost certainly have ecological effects. The ecological effects would include reduced arthropod biodiversity, together with decreased of numbers of beneficial insects. The number of required insecticide applications must certainly have negative ecological effects. There is little time for arthropod population recovery under most management plans for sweet corn, which is a different situation from the occasional insecticide application for non-Bt field corn. From the Syngenta study, the lack of recovery would also occur when using Bt sprays. In contrast, the obvious and large positive environmental effect of using Bt transgenic sweet corn is ours for the taking. It would be wrong not to adopt Bt sweet corn, if only for the positive environmental ramifications.

Even in light of the very encouraging Syngenta study, we do not fully understand all the nontarget effects that accompany the cultivation of large numbers of GM plants. Additional research is needed. But we also do not understand the full breadth of nontarget effects that organic or conventional farming practices impose. Although additional research is needed to study nontarget effects on arthropods, transgenic plants should be placed in the context of commonly accepted organic farming practices that include spraying Bt bacteria and conventional agriculture that includes chemical insecticide applications. And this is exactly what I like about the Candolfi et al. study.[13] It took a real-world, farm-level approach to study the effect of different crop protection scenarios. The headlines and some delimited lab studies have warned of potential negative environmental impacts of GM Bt crops on nontarget and beneficial insects. The Syngenta study demonstrates clearly that using Bt crops instead of conventional or organic agriculture is likely to increase the biodiversity of other biota found in agricultural fields. The story is not over—not by a long shot. But it is becoming clearer that, on the whole, transgenes to control damaging insects have tremendously positive, not negative, environmental effects.

Extensive Field-Scale Evaluations for Herbicide Tolerance

By far, the largest field experiments to test the environmental effects of GM plants are the U.K. field scale evaluations carried out in 2000–2003 by dozens of scientists at a cost of about $10 million. The purpose of the field-scale evaluations was to determine if transgenic herbicide tolerant crops, coupled with their appropriate herbicide, would decrease biodiversity on farms. While the focus seemed to be on GM plants, the studies were really about weeds and weed control. Specifically, how do weeds impact animal (mainly arthropod) biodiversity? In contrast with assessing biodiversity effects from Bt crops, which are relatively straightforward, the effect from herbicide tolerance in plants is more convoluted. Studying biodiversity side effects of Bt plants is certainly more logical than examining the effect from growing herbicide tolerant crops. Nonetheless, herbicide tolerance was chosen for this large effort.

Split-field experiments were performed in close to over 60 fields throughout the United Kingdom. Half of any field was planted with a conventional nontransgenic variety and the best weed management available (including herbicides), while the other half of the field was planted with an herbicide-tolerant variety and used the appropriate herbicide. The three crops tested were corn, canola, and beet. Sixty to seventy fields were used for each crop. Beet was engineered to be tolerant to glyphosate (Roundup) and canola and corn were engineered to be tolerant to glufosinate ammonium (Liberty herbicide); both of these herbicides control a broad spectrum of weeds, and the transgenic system allows over-the-top spraying of weed killer. The studies were published as 9 papers in a special issue of the *Philosophical Transactions of the Royal Society of London Biological Sciences* in October 2003.

Even though the authors of this suite of studies cautioned about overinterpreting the results of the experiments, the magnitude and scope of them leave little choice but for researchers and policy makers to closely consider their outcomes. And at the time of this writing, people are paying a lot of attention to the papers.

In general, over two growing seasons, weeds were controlled better in the transgenic sugar beet and canola plots compared to non-GM, but in corn, the opposite was observed. This converse result in corn is probably due to the effective use of the (now banned in Europe) herbicide atrazinie in pre-emergence applications in non-GM corn. In the 3 different crops, the diversity of weeds was not different between transgenic and nontransgenic types.[16] There were simply fewer weeds (except in GM corn). Furthermore, GM crops and appropriate herbicide application decreased weed recruitment from year to year, meaning that farmers using the GM system could expect continued success in controlling weeds from year to year.

Overall, the counts of arthropods living on soils followed the results of weed control. There were greater numbers of invertebrates in conventional beet and canola plots compared with their transgenic counterparts. The opposite was true in corn plots.[17] But it is interesting that differences in these ground-dwelling animals were greater among crops than between transgenic and nontransgenic types of the same crop. For arthropods living in the air, plant and soil, the same trend holds.[18] For example, there were fewer bees and butterflies in herbicide tolerant beet and canola plots, and this effect was related to the fact that fewer flowering weeds were left living. Once again, the differences among crops swamped transgenic differences within crop types. Regardless of transgenic state, there were approximately 40 times more bees and 5 times more butterflies in canola fields compared with cornfields; for corn, there were twice as many bees in transgenic plots compared to nontransgenic plots, and about 82% as many bees in transgenic versus nontransgenic canola fields.[18]

Several other aspects of biodiversity were studied. An entire chapter could be written on analyzing the data collected within fields and field margin effects. The bottom line of all the experiments is that weed control and choice of crop (along with insecticides) determine the kinds and amounts of insects, slugs, spiders, and other tiny animals that can be found in agricultural fields. There is no evidence that the herbicide tolerance transgene has any direct effect at all. Thus, the farm scale evaluations were more about management regimes than biotechnology. Similar effects could have been observed using herbicide-tolerant crop varieties obtained using conventional methods.

Farms are generally regarded as having the primary purpose of supplying *Homo sapiens* with food, and not butterfly food. As we saw with the monarch butterfly, killing milkweed, its sole food, affects its populations more than Bt corn pollen. The same has been shown in the British experiments. Weed densities do play an important role in the number and type of insects that are in the field.

At the time of this writing, it is still too early to determine the full societal effects of the research. Analysts have been pointing out some shortcomings of the studies and attempting to put the results in context, while GM opponents are taking advantage of the findings to suggest that GM plants are environmentally dangerous. Apparently, the popular media is happy to echo that sentiment. One article points out that "the UK's media has concluded that GM food is now a dead duck, which would make it very difficult for the government there to approve the commercial release of GM crops into the environment."[19(1419)] The title of one letter to *Nature Biotechnology* captures my overall opinion about the experiments' premise: "UK field-scale evaluations answer wrong questions."[20] In this letter, a who's who list of experts in agricultural biotechnology point out that GM plants are a side issue in weed control. The $10 million experiments could have just as easily examined the hoe.[20] One article authored by University of Oxford researcher R.P. Freckleton and colleagues notes that crop yields were not measured at all in the experiments.[21] Thus, potential risks to biodiversity are emphasized in the absence of potential benefits of increased crop yields. Furthermore, Freckleton et al. state that the most serious shortcoming of the research for public policy decisions is that it had no predictive component to forecast effects in 10, 20, or 50 years.[21]

So, how important is it to guarantee high biodiversity of spiders and slugs in crop fields? What about butterflies? In Britain? In India? On a scale of 1 to 10, with 10 being the highest, I venture to guess that in Britain, optimal crop yield and weed control would rate an 8, whereas having lots of butterflies would rank a 3. In India, 10 and 1, respectively. Britain is fairly unique in that farms are used for recreation.[20] Developing countries don't have the luxury to grow bugs in expense of food for starving people, and there, feeding people is more important than entertainment and recreation. But even in Britain, it is clear that if you want to maximize your yield of bees and butterflies, you should grow canola, not corn. The effect of transgenic variety and weed management is tiny compared to crop type.[18] Chassey et al. sum it up well: "[T]he ultimate question that should be asked is which agricultural technologies will maximize production while minimizing environmental impact in the broad sense. Herbicide-tolerant technology [GM and conventional] may be one of those rare technologies that improves both yield and product quality while reducing the environmental footprint of agriculture."[20(1429)]

Even though there seem to be few nontarget effects of Bt crops, the main thing that has been worrisome about growing large numbers of crops is that pests will become resistant to Bt toxins. If that happens, we would have failed to learn an important lesson from the chemistry school of hard knocks. Insects are genetic acrobats and quite versatile at evolving resistance to chemical insecticides. If Bt-resistant insects do emerge, not only would companies have lost a great investment, but organic farmers would also lose one of the few tools they possess for controlling insects on crops. Lessons learned from the evolution of resistance to chemical insecticides can teach us how to avoid errors in managing Bt resistance in GM plants.

8

Bt Resistance Management

Getting Off the Treadmill

The pesticide resistance treadmill is real. For as long as chemical insecticides have been applied, insects have evolved resistance to them, resulting in the loss of an effective chemistry to control the pest. If a farmer repeatedly sprays a chemical to kill a certain type of insect, it is likely that, after a number of seasons, the chemical will cease to be effective—the insects will have become adapted to it. Why would Bt endotoxin-producing plants be any different in this regard? At first blush, there would be no difference, but perhaps there could be special management tools that genetics give over chemistry.

Biology of Resistance

Recall the basic principles of population genetics. The rules are essentially the same for plant or insect populations (except in mating habits; more on that later). Molecules don't know whether they reside in plants or animals; DNA is DNA and proteins are proteins. Two factors of genetic change and microevolution are especially important in the following discussion: mutation and selection.

Mutation is the engine of genetic change in organisms. Let's say that a chemical insecticide inhibits the action of a certain essential enzyme (a protein) in insects. The insect will die when the chemical inhibits the enzyme. If the inhibition can be thwarted the insect lives because the enzyme can continue to be active in sustaining life. As in the case of all enzymes, they are encoded by genes. Since genes are continually subject to random mutations, albeit at a low frequency, say one out of a billion replications, by chance this enzyme-encoding gene could mutate to prohibit the insecticidal inhibition activities.

Selection is the sorter or sieve for favorable genetic changes. Selection, in this case conferred by a chemical insecticide, would "see" a favorable mutation that allows the insect to live and potentially reproduce. At the protein level, the mechanism could be a knocked-out active inhibition site on the enzyme's surface. The insecticide would no longer inhibit the enzyme, and the insect would live. If the insect reproduces with another insect of like resistance, a resistant population is born. The insecticide would no longer protect the crop from insect damage as the resistant population grows and spreads. The insect has won this battle in the evolutionary arms race. This is one accepted model of insecticide resistance—one that has been played out countless times on America's farms in the past few decades.

By all accounts, there is no reason to believe that insect biotypes would evolve resistance differently in response to Bt transgenic crops. In Bt resistance management, we worry most about the receptor on the insect midgut cells, which is the most likely target for mutation and selection for resistance. A Bt receptor protein can be coded for by one or more genes, which increases potential mutational targets. There are at least two distinct possibilities for altering the receptor resulting in resistance to Bt endotoxin. The first possibility is that the receptor would not be produced. A frameshift mutation, in which a nucleotide is added or subtracted in the coding frame of the receptor gene, would cause the receptor to not be produced. If the receptor is not produced, Bt toxin cannot bind to gut cells, which results in zero toxicity by Bt. The second possibility is that the receptor would still be produced, but not bind Bt toxin (figure 8.1). This latter case would arise if a point mutation changed the Bt binding site with an amino acid substitution. Either of these possibilities is feasible, but the second example is a more likely scenario. In the second case, Bt binding could be decreased instead of eliminated, thereby altering the toxicity of Bt.

These models are not just hypothetical, as Bt resistance has already occurred in more than one insect. Bt-resistant insect populations have been created in the laboratory, and even more important, they have appeared on the farm (not laboratory escapees). One insect example was described previously—the diamondback moth (*Plutella xylostella*) is an important

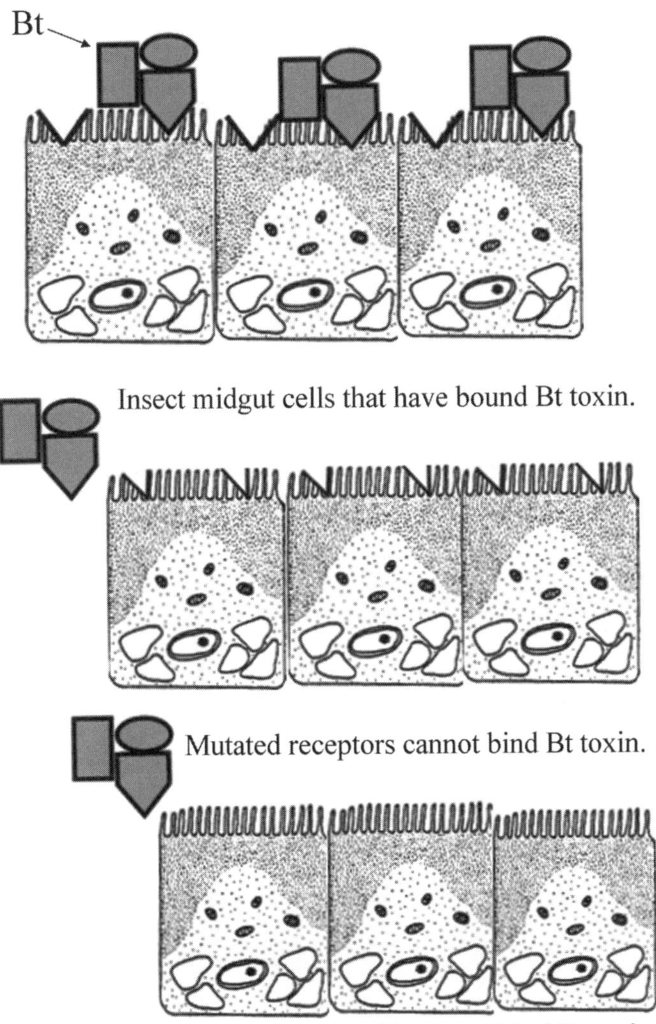

Figure 8.1. Different mutations leading to Bt resistance. In the top panel, receptors are intact and able to bind Bt endotoxin proteins to midgut cells. Result: insects will die. The middle panel shows receptors that are altered by a mutation. Result: receptors can no longer bind Bt and insect lives. The bottom panel depicts a situation in which the receptor is missing as the result of a mutation. Result: Bt is not bound and insect lives.

pest on mustard crops (broccoli, cauliflower, cabbage, etc.). Bt-resistant biotypes in American fields have been discovered in Hawaii, Florida, and New York. Populations of Bt-resistant diamondback moths have also been found in Central America and in central and southeastern Asia in multiple locations.[1] Resistant biotypes were selected by Bt sprays, such as the types used by organic farmers, not by Bt-transgenic plants. Apparently, it is not difficult for this pest insect to become resistant to Bt-based insecticides. It is important to note, too, that there are different genes resident in certain insects that can confer resistance. There is not simply one target gene. The examples shown in figure 8.1 are probably the most simplistic scenarios of Bt resistance. These various resistant species discovered to date, and even populations of the same species, may not have the same resistance mechanisms (hence, genes/alleles). One more thing: insects that are resistant to one Bt toxin might be cross-resistant to another toxin. For instance, lab populations of the tobacco budworm (*Heliothis virescens*) selected for resistance to Bt Cry1Ac were also found to be resistant to Cry1Aa, Cry1Ab, and Cry1F insecticidal proteins.[2] It is clear that the problem of managing resistance to Bt in transgenic crops is a pressing and complex issue. Insects have more than one way to skin the proverbial cat.

Bt Resistance: So What?

What are the ramifications of insect resistance to Bt? Would it matter to farmers or the environment if Bt genes lost their effectiveness in insect control? After all, aren't scientists working to discover new insecticidal genes with novel modes of action; genes that code for proteins that are dissimilar to Bt toxins? While scientists in academia and industry are busy finding new insect resistance genes that might be useful in GM plants, no one wants to see the current family of Bt *cry1* genes lose their effectiveness anytime soon. Companies have invested considerable resources in developing Cry-producing transgenic insecticidal plants. They would like to recoup their investment as long as possible through seed sales of transgenic crops. Nearly everyone views Bt genes and proteins as global assets whose effectiveness should be maintained as long as possible. No one wants to see these insecticidal proteins devalued in any way.

There are good biological reasons to protect the viability of Bt toxins. Certain insects might be optimally controlled using a certain Bt protein.

Perhaps no substitute protein-based toxin with a closely matching specificity exists in nature. Then there is the cross-resistance problem. There is no guarantee that, if a Cry1Ab-resistant insect population appeared, using plants that produced Bt Cry1F would help control this population. The insects might be cross-resistant to Cry1F as well. And if the change between one Bt toxin and another is a small step, say, Bt Cry1Ac to Cry1F, then a more robust resistance to both could be selected for; cross-resistance to more than one Bt toxin would be more difficult to overcome that for single resistance. Once resistant insect biotypes are selected for, there is no evidence that the resistant genes would be purged from a population; resistance could be permanent.

It has been argued that organic farmers would be the group to suffer most if insects gain resistance to Bt. After all, Bt sprays are some of the few insecticides their philosophy and commodity guidelines allow them to use. If insects are resistant to Bt toxins, how could organic farmers combat insects?

No one would like to see the chemical insecticide treadmill reproduced in transgenic insecticidal crops, but most mathematical models suggest that the best we can hope for is to delay the time for resistance to surface. Although resistance to Bt may be inevitable, most people agree that we should strive to delay its arrival. The time scale I most often hear batted around is 20 years. But I've heard that now for 8 years, so does that mean that the first generation of Bt crops have 12 years left on the time-to-resistance clock? Most of the prognostications are proffered by computer modelers whose predictions are only as good as their assumptions and models. In such a biological system with many unknown parameters, computer models have limited hope of absolute accuracy. Certainly, 20 years is long enough for companies to obtain a decent return on their investment and develop new GM insect resistance crops, but I doubt that organic farmers would be pleased about any degree of Bt resistance that surfaces, no matter what the timing. It is noteworthy that Bt-resistant diamondback moths around the world arose as a result of organic farmers spraying Bt bacteria on *Brassica* crops, not as a result of GM plants (there are no commercially grown Bt *Brassica* crops yet, although there is Bt *Brassica* in development in India, and commercial Bt canola candidate has been field tested in the United States and elsewhere). The fact that organic farmers have already contributed to Bt resistance of diamondback moth has prevented most companies from considering it to be feasible to release Bt transgenic *Brassica* crops.

Delaying Resistance

Many academic and government scientists have been active in examining ways to delay Bt resistance. In fact, the Environmental Protection Agency (EPA) explicitly regulates the cultivation of Bt transgenic crops and is the lead regulatory agency mandated to oversee Bt-resistance management plans in the United States. In contrast, Bt sprays (from an organic production perspective) and Bt-resistance management are not regulated to the same extent as GM plants. Much the same can be said for conventional chemical insecticides.

There are essentially two primary strategies to delay Bt resistance: high dose and refugia. Two secondary strategies are pyramiding (gene stacking) and crop rotation. There have been other approaches discussed among experts, but the scientific consensus is that these are our best bets. The EPA has mandated usage of a high-dose/refugia combination strategy. In addition, companies are beginning to stack multiple insect-resistance transgenes together for broader control of pest insects, so it is feasible that all three strategies will be used in concert to delay resistance to Bt.

High Dose

The high-dose strategy is a sledgehammer. If enough Bt toxin protein is produced by the plant and delivered into the gut, target insects will not stand an evolving chance. I've heard a couple of academic scientists purport that the high-dose strategy would be wholly sufficient to totally prevent resistance from occurring. Most scientists, though, recognize that it would simply be a matter of time before a couple insects lacking Bt receptors will be born and reared in the same proximity to then mate and parent a new population of resistant insects.

The high-dose strategy alone would merely strongly select for resistant insect biotypes. The EPA has defined a high dosage as 25 times the amount of Bt protein required to normally kill the target susceptible insect (LD_{50}). Because little Bt toxin is required for lethality for certain target insects, the required high dose is really not very high, thereby making it feasible to impart. Besides, farmers and vendors of the transgenic plants desire high dosage for maximal insect control anyway. Low Bt dosages could allow field-level crop damage if some target insects were allowed to survive. No one has argued against the necessity of implementing a high-dose strategy, except to state that continually producing a toxin in all tissues of a plant might lead to side effects, such as unnecessary exposure to beneficial insects, leading to unintended consequences. But as we saw in the previous chapter, there are no data in support of this viewpoint at this time.

Refugia

The refugia strategy uses spatial patterns of Bt and non-Bt crop deployment to delay the onset of resistance. Reproductive islands of Bt-susceptible individuals can flourish if non-Bt crops are grown adjacent to transgenic crops. If a resistant individual insect does emerge and lives to mate in the Bt field, her mate choice would consist of an overwhelming majority of susceptible insects, thus producing no resistant offspring. The refuge plan mandated by the EPA specifies that farmers not plant all their land with Bt crops but sow seeds of non-Bt plants of the same species on some parcels. If the farmer wants to spray chemical insecticides, he must keep a bigger refuge. For example, the EPA requires growers of Bt cotton to plant 25% of their crop with non-transgenic cotton that can be treated with chemical insecticides (non-Bt insecticides). The grower can plant 4% of the crop with non-transgenic cotton in which no insecticides will be applied; these are sacrificial plants.

Most of the Bt resistance traits in insects characterized thus far are completely recessive. The completely recessive resistance case is the theoretical expectation predicted by receptor biology. Bt receptors in insects are digestive enzymes (such as an aminopeptidase) and cell adhesion proteins (such as a cadherin) that insects obviously use for other purposes that have nothing to do with binding Bt. An overwhelming majority of individual insects in a target species produce completely functional receptors that would render them susceptible to Bt. So, the typical target insect that had functional receptors would die if it ingested Bt toxin. In the case it had the typical two copies (alleles) of the gene encoding a susceptible Bt toxin receptor protein, it would be a homozygous susceptible (*SS*) individual, where *S* is a dominant allele. Such an insect having this genotype would be killed by Bt toxin. The Bt toxin would bind to the Bt receptor, then punch holes in the insect's guts. If a mutation appeared in one of the alleles, not as many functional (susceptible) receptors would be produced, but the insect midguts would still bind Bt, resulting in death. The genotype in this case would be *Ss*, and the individual insect would be heterozygous for the Bt receptor gene. Because insects with this genotype would be killed by Bt toxin if it were present, just like homozygous *SS* individuals, they would, of course, not live to reproduce. Note that *s* signifies the recessive allele that allows Bt resistance and survival when Bt is ingested. Only individuals with the genotype *ss*, homozygous recessives, would be resistant to Bt toxin and live to reproduce.

For the purpose of this illustration, let's assume the resistance allele (*s*) is extremely rare in a population of insects. And indeed, in real life, this is the case. If insects are challenged with Bt, then only those insects that have no functional receptors for Bt (*ss*) would live to reproduce. That the trait is entirely (or nearly entirely) recessive is the whole reason the refugia strategy can work.

Because resistant individuals are rare, let's say one in a million insects, we don't have to worry about them very much as potential mates for each other. But if resistant individuals are one in a million, then resistant alleles are only one in 1000. The reasoning: frequency of $s = 0.001$. The probability of two resistant alleles (*s*) coming together in one insect is $0.001 \times 0.001 = 0.000001$: one in a million as predicted by the Hardy-Weinberg equilibrium (chapter 4).

Let's say that spontaneous mutations occur in insects resident in Bt and non-Bt fields with 75% of the land under cultivation of a Bt crop and 25% of the land under cultivation of a non-Bt crop. Only those insects with genotype *ss* will survive in the Bt fields. But these insects will be extraordinarily rare—one in a million. But the one in a million will be looking for a mate, nonetheless. It will fly around for some time in its Bt field to find that no mate is likely to be available (all dead), so it travels into the adjacent field with hopes of getting lucky. There will be numerous mate choices from the 25% refuge field. The overwhelming majority of potential mates will not have any mutation (*SS*)—to be exact, 999,000 out of a million. The progeny of this mating would all be heterozygous individuals (*Ss*) and would be killed if they happened to ingest Bt toxin. Most resistant insects will likely find this type of mate. Of the million choices of mates the *ss* (resistant) individual from the Bt field could mate with 999 *Ss* individuals (of 1 million). Half of the offspring of this union would be *Ss* and half would be *ss*. Only one in a million insects from the refuge field would be available to mate with the *ss* individual to produce all resistant insects.

Although it is impossible to guarantee that the refugia strategy will always succeed, it should effectively delay the onset of resistance, especially when it is coupled with the high-dose strategy. Confidence in these strategies presumes that a common mutation conferring resistance is rare and does not appear in numerous individual insects in Bt fields. As the mutation becomes more common, it will be increasingly (exponentially) easier for resistant insects (*ss*) to find one another in the close proximity for mating. There is something to be said for convenience. Once homozygous (*ss*) individuals mate, only homozygous (*ss*) offspring will be produced, and resistance would be fixed in the insect population rather quickly, which will allow the populations of resistant biotypes to increase. The refugia strategy will not be effective if resistance genes are not recessive. However, there are additional things that can be done to delay resistance even further.

Crop Rotation

As part of any IPM strategy, crop rotation is essential. If a farmer plants cotton year after year in the same field, cotton pest insect problems grow increasingly worse year after year. Crop rotation moves the location of insects' preferred food source annually by alternating other crops in off years by planting cotton in a specific field only once every second, third, or fourth year. The pest insect populations are thus forced to migrate because their food source becomes unpredictable in a single location. This farming practice also helps decrease the probability that insects with resistance alleles (*s*) will be able to find mates possessing resistance alleles. The resistant population would have little chance of becoming established because the Bt selection pressure would be removed some years. Here is where organic farmers that selected Bt-resistant diamondback moths erred. They planted *Brassica* crops year after year and sprayed Bt insecticidal bacteria year after year, providing steady selection pressure allowing Bt-resistance alleles to get a foothold in the insect population. The resistant alleles could then become fixed in a population. Responsible farmers practice crop rotation and don't use Bt crops (or sprays) year after year.

Pyramiding

This kind of pyramid does not refer to a scam to make lots of money from unsuspecting victims. Pyramiding, or gene stacking, is genetically engineering two or more insecticidal genes encoding proteins that have different modes of action. The reasoning is that if insects could, by chance, survive one type of Bt gene, then another insecticidal transgenic product could very well kill them, assuming no cross-resistance is present. It would be much more improbable for the insect to evolve two steps of resistance simultaneously in one generation. If we started with probabilities of one in a million for both types of resistance genes, then now it would take one in a trillion for both types of resistance to arise at the same time in the same insect. Companies are progressing toward placing pyramids into commercial crop plants. For example, Monsanto has released a cotton variety with two Bt toxin proteins (Bollgard II). Research is also underway to discover genes other than those that encode Bt endotoxin proteins. The potential problem of using two Bt genes is that the modes of action could be too similar to one another, and there could then be cross-resistance among Bt toxins, as discussed earlier. However, since a potential second non-Bt insect resistance gene for pyramiding has largely evaded researchers, a second Bt gene is better than nothing. In fact, an experimental two-Bt gene system in transgenic plants has been proven effective in delaying the evolution of resistance.[3]

New Insect Resistance Genes

Bacillus thuringiensis possesses the greatest diversity of insecticidal proteins of all known species of organisms on earth. To date, hundreds of Bt strains have been discovered, and several different classes of crystal toxin proteins have been described on the basis of the types (orders) of insects they kill.

The best choice of an insect resistance gene for pyramiding with a Bt gene in cotton (*cry1Ac* was the first gene) was a second Bt. There are several reasons this was a straightforward choice for Monsanto, which developed Bt cotton sold under the name Bollgard II. First, there are many Bt genes that have been cloned, sequenced, and completely characterized. Second, farmers, regulatory officials, and, to a lesser degree, consumers are familiar and comfortable with Bt toxins expressed in plants. Third, there is no other obvious choice that yields equivalent efficacy and specificity of killing the host insect compared with another Bt. The Bollgard II cotton is the second-generation insect-resistant cotton that has a Bt *cry2A* gene coupled with the initial *cry1Ac* gene. Two Bt genes should delay the evolution of resistant insect biotypes because no cross-resistance has been noted between Cry1Ac and Cry2A. In addition, this pyramid should also increase the efficacy of crop protection against insects. If Monsanto's Bollgard II cotton is as successful as predicted, then we should expect Monsanto and other companies to introduce additional transgenic crop plants with multiple Bt genes. Eventually, however, we must look beyond Bt genes for insecticidal pyramiding candidates.

Various companies have discovered non-Bt sources of protein-derived protection against insects, which are discussed further elsewhere.[4] Monsanto has worked on cholesterol oxidase from filtrate of *Streptomyces* cultures, which has been found to be highly toxic against the boll weevil. So a cholesterol oxidase transgene might be useful in cotton. However, this protein also has activity against a corn rootworm and other corn pests. The mode of action is the lysis of midgut cells. Syngenta scientists have published papers on Bt vegetative insecticidal proteins, a different class of insecticidal compounds from the Bt endotoxin proteins discussed throughout this book. Similar to cholesterol oxidase, vegetative proteins also kill insects by breaking open insect midgut cells. The Vip3A insecticidal protein has been shown to be toxic to black cutworm, fall armyworm, beet armyworm, tobacco budworm, and corn earworm. These species represent a broad range of insects that damage corn. Toxins have been discovered from the bacterium *Photorhabdus luminescens*, which are found in certain nematodes. Dow AgroSciences has examined these toxins that are toxic to a number of different insects. In addition to these bioactive proteins, other proteins such as chitinases, proteinase inhibitors, and lectins have been assessed for their control of insects. A potential marketing advantage of most of these protein-derived insect control agents is that they originate from plants and might be more acceptable to consumers than transgenes originating from bacteria. The downside is that they don't work as well as Bt toxins. However, a proteinase inhibitor from a species of *Brassica* was recently found to be as effective as a Bt endotoxin in killing certain caterpillars.[5]

Ecological Effects of Bt Resistance Management

I have included a chapter on Bt resistance management in this book because it is the most timely and real environmental risk associated with today's transgenic crops. But there might be a tradeoff between delaying resistance (good) and nontarget effects (bad). New and novel gene discovery for proteins that kill insects will lead to transgenic crops that kill a wider variety of insects than Bt. A priori, broad-spectrum action is initially advantageous from the company and farmer perspective. Imagine the ability to produce one or two proteins in plants that prevent any herbivore candidates from eating the crop—a *deathstar* gene. That would be fabulous! Everyone would be happy (except for the nontarget insect, that is). We have seen that because Bt toxins are so very specific for target insects, their side effects are few. This is a direct relationship. It is foreseeable that more types of nontarget effects would be observed in fields with more broad-spectrum transgenic crops. These new transgenic crops will have to be assessed on a case-by-case basis, just as Bt crops are today. As they come closer to coming on-line, our assessment methodologies should have also matured.

I believe that concurrently with availability of different modes of transgenic insect control, more precise deployment of the expression of resistance genes will also emerge. One of these modes will be inducible expression. Currently, transgenes are expressed everywhere and all the time in a plant (constitutive expression). This is the simplest way, but it may not be the best. A more precise and strategic expression pattern would be desirable to decrease potential nontarget effects and to also delay the evolution of resistant alleles. For example, while root exudates of Bt toxins have been found not to be biologically important to nontarget arthropods, one might easily imagine a broad-spectrum insecticidal protein having a biological effect in soil and water (undesirable exposure effects). Because most herbivorous pest insects feed on leaves, why squander transgene expression in roots where potential unnecessary nontarget effects could occur? Likewise, current insecticidal transgenic crops produce their protective proteins at times they are not needed, such as when insect pests are not present. Imagine transgenic toxins being induced by the plant just when insects begin to feed. The two improvements just described would require, first, tissue-specific promoters that only drive gene expression in leaves (or fruit or root, as needed) and, second, wound-inducible promoters that require slight amounts of defoliation by insects to induce gene expression. These types of promoters are being discovered and characterized and will no doubt play an important role in more precise expression of insect resistance (and other types of) transgenes. One group of researchers has already produced plants that express a Bt gene upon specific induction.[6] Unfortunately, the expression of Bt was too low to be commercially useful, but this study demonstrates that an inducible strategy might be feasible.

Perspectives

Clearly, engineered insect resistance can make agriculture more sustainable by decreasing reliance on chemical pesticides. GM practices also fit well with IPM. But not everyone is happy with this arrangement.

Agriculture, whether conventional or organic, is not particularly environmentally friendly by anyone's estimate. Agriculture drastically changes ecosystems in any format it is practiced. As of this writing, no one has proposed a more feasible way to feed the earth's growing population than large-scale farming utilizing monocultures. Organic farmers claim to have a viable alternative, but the yield data in the best of circumstances do not support their thesis (more on this in another chapter). A couple years ago I heard a scientist advocate the practice of the farming of and the subsequent eating of insects as an alternative plan for providing calories for humans; the term he used, I think, was "mini-farms." I don't think people will generally embrace the products of such a mini-farm.

The whole purpose of farming is to minimize local aspects of biodiversity for the sake of obtaining maximum seed (or fruit) yield of a single type of plant to feed only humans. Initiating and maintaining agricultural fields converts whole ecosystems from what they were naturally to farms. But this system has been the best one people have come up with in the past 10,000 years for providing large amounts of food for a reasonable cost, even though there is considerable room for improvement in environmental stewardship. Increased plant biodiversity in fields is called weed problems. Increased insect and mite biomass and biodiversity in agriculture is commonly called herbivory. Increased fungal and bacterial biodiversity readily translates to plant disease problems. There are examples, of course, where these generalizations do not hold. For example, plants grow better when they have mycorrhizal symbionts in their roots for improved nutrient uptake. As we've seen, there are beneficial insects, too. Certain conventional agricultural practices such methyl bromide fumigation, which sterilizes soil, is dangerous, ecologically unfriendly, and being phased out by the Environmental Protection Agency. Biofumigation and biotechnology can help replace methyl bromide. My point here is that conventional agriculture, while not perfect, is the best we've invented as a system.

If we are constrained to using monocultures to obtain large yields (and I think we are), then what practices or types of plants can minimize further negative environmental impacts? How can technology help us be better environmental stewards? As we've also seen, Bt plants are generally more environmentally friendly than conventional insecticidal sprays. Agreeing to implement strategies to increase the longevity of Bt production in transgenic plants is good stewardship, while denying the biotechnology altogether damns us to a world with no reasonable chance of improving yields or environmental sustainability. Organic farmers would like to banish Bt transgenic plant biotechnology, and, in that regard, theirs is a parochial view that favors organic farmer economics, but would drastically decrease yield and the global food production. In addition, there are no data to suggest that Bt crops will, in fact, diminish the ability of organic farmers to practice their preferred agricultural traditions. Since organic farmers grow precious little of the food that Americans (or any other group of people) eat (only 3% of the fruits and vegetables sold in the United States are organically grown), their views should hold proportional weight to their productivity. Finally, it seems odd to me that there is little regulation of pest treatment in organic crops (e.g., frequency and spatial control of Bt sprays) because their insect control practices (e.g., the overuse of Bt sprays) has caused specific negative environmental impacts (e.g., Bt-resistant insects).

Like many issues in GM plant commercialization and agriculture, the real biological concerns revolve around farming itself and not around environmental impacts per se. The majority of the environmental impacts of Bt crops are overwhelmingly positive. The planet is a better and safer place because of Bt transgenic plants. Although field-level nontarget effects of Bt genes are practically absent, there might be side effects when broad-spectrum insecticidal proteins are expressed in plants. That bridge will be crossed when we come to it, and when we have the regulatory and scientific structure to allow us to cross it efficiently.

9

Swap Meet from Heck

Trading Sequences between Viruses and Transgenes

One success story in the plant biotech world has been the creation of GM virus-resistant crops. Viral diseases, while often not as debilitating to crops as weeds, insects, or even fungal diseases and nematodes, are significant problems to certain crops, such as papaya and squash. In addition, viral diseases have traditionally been hard to control. Because viruses are not organisms per se, one cannot "kill" them directly—the best that most farmers can hope for is to eliminate the vector, or, at least, to decrease vector populations. The vector is typically an insect pest in its own right that not only eats part of the plant, but also transfers virus and other disease-causing agents from plant to plant. Just think about the last time you suffered the most prevalent human viral disease, the common cold. There was nothing you could do to get cured other than to wait for the infection to run its course. When we contract a cold (and then usually subsequently afterward), we wash our hands more frequently to reduce the spread of the virus—we decrease the vector. The same principle exists in controlling other human viral diseases such as malaria or West Nile virus. The most effective prevention is to kill the mosquitoes that vector these viruses.

Although some plant viruses are DNA based, the majority of the most troubling ones are RNA viruses. The same holds true for those affecting people, such as HIV. RNA viruses consist of a single strand of RNA, associated proteins, and a few simple ancillary biological molecules (figure 9.1). Because the virus lacks the molecular mechanisms needed to replicate, it makes its living by commandeering the plant's enzymes and cellular machinery to reproduce, which is often sufficient to make the plant sick. Plant

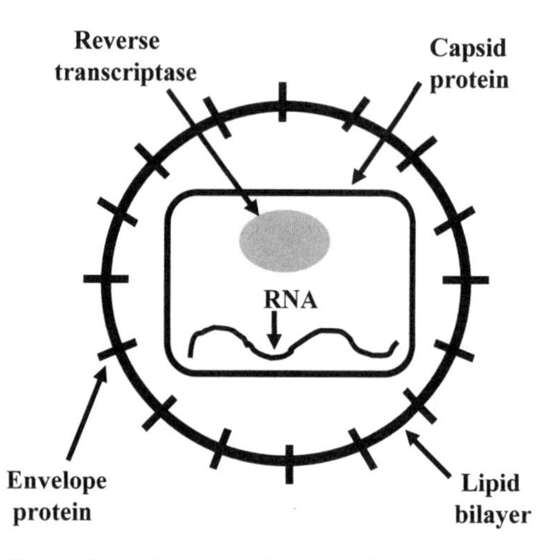

Figure 9.1. Diagram showing the simple features of an RNA virus. (Figure by Mentewab Ayalew and Reginald Millwood.)

viruses are the subject of active study in plant biology, from their molecular biology to the epidemiology of spread and infection.

Let's take a closer look at controlling insect vectors as a means to prevent viral infection. Viruses are most often transmitted from plant to plant by insects that have piercing-sucking mouthparts, such as aphids or other insects, such as thrips, that wound plant tissues. These insects extract nourishment from the plant while keeping plant organs and structures largely intact. Feeding habits of these insects contrast with caterpillars that remove whole leaves or eat holes in leaves—defoliators. Insects' piercing-sucking mouthparts act as tiny syringes to inject viruses from plant to plant, placing the virus particles directly in the stream of plant vasculature. Similar to defoliating insects, viral vectors are most frequently controlled by applying chemical insecticides, which have potentially significant negative ecological impacts. As an aside, recent research has shown that plants do have a natural method of combating viruses by silencing the expression of viral genes. The mechanism by which they do so is a very hot topic in plant biology research called virus-induced gene silencing and post-transcriptional gene silencing.[1] One key component of gene silencing is small, double-stranded RNA, which causes a viral transcript to not be translated or replicated and, in fact, targets it for degradation. So, while some plants have natural viral resistance through this and other mechanisms, viral diseases are nonetheless devastating to certain species. Transgenic plants have proven to be very effective in conferring virus resistance.

Virus-Resistant Plants

A case in point of a debilitating viral disease on crops is ring spot papaya virus. It has threatened the worldwide production of papaya on subsistence farms throughout the tropics. Papaya is also an important crop in Hawaii, birthplace of Dennis Gonsalves, who is now a scientist in Hawaii at the USDA. Dennis Gonsalves's laboratory engineered papaya to be resistant to ringspot papaya virus by introducing a transgene that codes for a coat protein, a very useful virus control strategy that has been used more often than any other in transgenic plants.[2] Some people contend that GM virus-resistant papaya will save the papaya industry in Hawaii. In the 1940s, ringspot virus was discovered in Oahu, where it destroyed papaya cultivation; papaya moved to the big island of Hawaii, where the virus was absent. The virus was discovered there in 1992 and began to decimate papaya production. By that time GM papaya for virus resistance had been developed and was being tested.[3] As of 2002, the transgenic plants had been tested and grown in Hawaii successfully for 7 years with the latter part of the period in commercial production. All data indicate complete protection against virus and fruit yields three times higher than the industry average (figure 9.2).

Figure 9.2. A healthy papaya plant (left) and one infected with papaya ring spot virus (right). (Photos by Dennis Gonsalves.)

The other notable transgenic viral-resistant crop engineered to date is squash.[4] Limited acreage of transgenic virus-resistant squash by the company Asgrow has been grown in the United States. In this case, coat protein-encoding transgenes from the cucumber mosaic virus, watermelon mosaic virus 2, or zucchini yellow mosaic virus were used.

There are other transgenic approaches that could be effectively deployed other than expressing viral coat protein genes in plants. Roger Beachy, one of the earliest experimenters with genetic engineering, pioneered the approach expressing an antisense viral gene which would yield a backward transcript that would bind an essential viral gene, thereby preventing its translation.[5] Other methods that have been investigated are programmed cell death of the infected plant cells to restrict the movement of the virus and expression of foreign genes that could interfere directly with the virus or its genome.

So far, the most effective transgenes used to produce virus-resistant plants utilize parts of the virus genome itself. This approach presents a unique risk: recombination of the transgene with parts of the virus to create a new novel virus caused by GM plants.

Recombination

There are essentially two categories of ecological risks associated with virus-resistant transgenic plants. The first is not unique and has been discussed earlier, but using other examples—gene flow and persistence of fitness-enhancing genes from crops to wild species. Viral resistance genes could conceivably enhance the fitness and competitive ability of wild plants, thereby producing more troublesome weeds. Not much is known about virus infections in wild plants, and therefore more basic research is needed to understand the biological background. So, other than to say that more research is needed, and since this risk is not specifically related to transgenic plants, we will not dwell on it further in this chapter. The potentially more serious and unique risk revolves around virus recombination. If viruses recombined with transgenes, the new viruses could be more virulent and infect a wider range of hosts.

Before discussing transgenic plant-specific recombination risks, let's explore what is known about natural RNA virus recombination. Without any human intervention, viruses are masters of genetic change and fluidity: "RNA viruses deserve their reputation as Nature's swiftest evolvers. Their high rates of mutation and replication allow them to move through sequence space at a pace that often makes their DNA-based hosts' evolution look glacial by comparison."[6(2535)] Part of their high rate of evolution is the result of high mutation frequency. The other half of the equation is recombination; viruses swap nucleotide sequences between strains and within and among virus types. They can even recombine nucleotide sequences with their hosts. There are two types of recombination: One is homologous recombination, in which sequences and regions are similar, and the other is nonhomologous recombination, in which the swapped sequences can be quite different. In either case, the sequences serve to potentially increase its virulence or success. Homologous recombination would typically occur between viral strains or transgene and virus, whereas nonhomologous recombination occurs between different kinds of viruses. Because nonhomologous recombination has not been shown between viruses and transgenic plants, the remainder of this chapter focuses on three questions involving homologous recombination: (1) could viruses and virus-based transgenes feasibly recombine to form a different virus? Has this been demonstrated? (2) Can recombination happen in the field? (3) If it does happen, would it result in the viral plague from heck? How likely is it that a recombination event would create a new virus that is even slightly more infective than its predecessor or that would cause the virus to broaden its host range to other plants?

Viral Recombination under Experimental Conditions

There is no question that viruses can recombine with the transgenes that originally come from virus (e.g., viral coat protein genes). There have been several different combinations of transgenes, plants, and viruses that have been tested for homologous recombination. The earliest experiments all used similar approaches. An experimentally crippled virus that could not infect and replicate in plants was inserted into transgenic plant cells. The transgene complemented the viral genomic deficiency. After recombination had occurred, the new recombinant virus could be detected upon strong selection pressure. The most efficient screening system consisted of looking for a new functioning virus that could only exist if effective homologous recombination had occurred. We will examine a couple of sophisticated early (early 1990s) experiments. The cauliflower mosaic virus (CaMV) requires a gene coding for a transcriptional transactivator (gene IV) in order to replicate for systemic infection (i.e., outside of the inoculation spot). Researchers infected transgenic plants with a mutant of CaMV that was deficient in gene IV. Experiments were performed with canola, one of its natural hosts, as well as tobacco, a plant that is outside of the CaMV's optimal natural host range. Transgenic canola was produced that expressed a functional gene IV, and the virus was subject to strong selection.[7] In this case, a new recombinant virus able to travel inside the plant was recovered that had obtained gene IV via homologous recombination (figure 9.3). In a related study, a mutant CaMV strain that could not infect tobacco was able to infect transgenic tobacco (that had gene IV) because of the virus recombining with the transgene RNA.[8] In both of these cases strong artificial selection pressure was imposed to find the recombinant viruses. Subsequent experiments from various teams of researchers have demonstrated that even weak selection pressure is sufficient to recover recombinant viruses from transgenic plants.[9] The bottom line is that in experimental systems, recombination will occur between transgenic plants expressing viral genes and viruses.

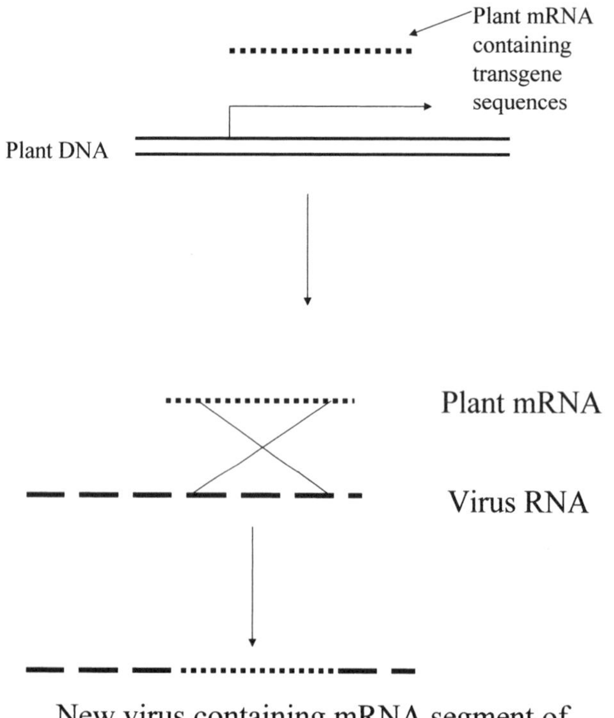

Figure 9.3. Viral RNA can recombine with plant RNA, including that which includes transgenic RNA sources originating from the virus itself.

Recombination in the Real World

Because viruses naturally recombine in the field, it is expected that viral recombination with transgenic plants would also occur in the field. However, due to obvious logistical and resource constraints, there have been few studies performed to quantify recombination in a field-based system. One of the few studies was undertaken over a 6-year period. Transgenic potatoes that expressed either coat protein or viral replicase genes for potato leafroll virus were grown in the field, and viruses were sampled from the potatoes to assay for recombination.[10] There were no significant changes in any viral characters that would indicate the virus had recombined with transgenes to create a more menacing virus during the 6-year period. But because wildtype viruses were found in the field naturally, there was no obvious fitness enhancement that the transgenic sequences could confer. And I think this point is well worth considering. Under these conditions, what would be the consequence if recombination did occur, if the sequences derived from the native virus were identical to the one being used in the transgenic plant? It would be like swapping one black shoe for another black shoe—nothing would be out of place, and no advantage would be gained by the homologous recombination.

A Supervirus Is Not Likely to Arise from Recombination with Transgenic Plants

We see a thread of common themes when we examine experimental results designed to assess the various kinds of risks of GM plants. We are compelled to ask a few recurring questions. The first is the "what if?" followed by "so what?" The latter is the question of consequence when we compare a transgenic plant with the status quo in the real world of agriculture or nature. Will the transgene–crop combination cause significantly greater risks than a non-GM counterpart system[8]? And the final question is one of mitigating risks: what can we do in either the engineering or deployment of a GM plant in an agricultural system to maximize benefits and minimize risks?

People are concerned about viruses recombining with transgenic plant sequences in the field (the "what if" question). One helpful related question is, do viruses recombine in nature—nonhomologous recombination to form new and more virulent viruses? There is precedent for viruses undergoing homologous recombination to form hybrid viruses.[11] But Rubio et al. raise another interesting question: we know that RNA viruses have very rapid rate of evolution, but if recombination is so important, then why don't we find hybrid viruses in nature very frequently?[12] It seems that instead of the new recombinant viruses playing the part of superviruses, they are generally unable to compete with existing viruses that already successfully infect plants. How would transgenic plants be any different from the case of virus–virus recombination, which tends to produce novel but crippled recombinants?

There are several reasons to support the hypothesis that transgenic plants expressing coat protein genes or other transgenes originating from virus will actually reduce virus recombination and slow the speed of virus evolution. GM plants could actually decrease virus recombination risks. First, as stated above, when a virus coat protein gene is engineered into a plant to protect it from the virus, no new genetic information is introduced into the system. The genetic information contained in the transgene is contained entirely in the viral pest of interest. Continual recombination with plant-produced coat protein genes of the original virus could slow virus evolution that might otherwise enhance the viral-encoded coat protein. The second reason that transgenic plants might aid in delaying evolution in viruses is that the plant would actually suppress viral loads in the plant. There should be fewer viral copies in the plant where recombination and evolution take place, resulting in fewer opportunities for recombination. In contrast, there are more copies of the transgenic-borne transcript—say, more coat protein transcripts available for recombination with virus. But, obviously, the limiting factor for recombination would be lack of the virus itself in that circumstance. Third, and related to the second reason, in addition to a reduction in viral load in the plant, viral replication has been shown to be decreased in transgenic plants.[12] Since replication is the golden moment for recombination, the opportunity for recombination should be significantly less in transgenic plants as a rule.

What Next?

Because virus-resistant transgenic papaya and squash are already deregulated, it is still imperative that additional field data be collected in large-scale experiments on homologous viral recombination. There needs to be more funding for such studies. Because viruses are essentially single-stranded RNA molecules, time-course studies can be performed to gene bank viruses to study in the nucleotide sequence changes over time. In addition, it could be determined whether having transgenic plants in the equation makes a difference in evolution rates.

To mitigate potential risks, Richard Allison and colleagues analyzed engineering techniques that could minimize recombination and proposed three recommendations to that end.[13] First, they recommended that transgene candidates should exclude known virus replication initiation sequences. They admitted that we did not have a complete list of these sequences (in 1996), but we do have more information today to help guide such decisions. The second recommendation was that the transgene should consist of the smallest piece of nucleic acid sequence to do the job of conferring resistance. Smaller is better because fewer targets would be available for recombination. Finally, the expression level of the transgene should be minimized. This last recommendation runs counter to most transgenic systems. Minimizing expression levels would decrease the number of transgenic, virus-based transcripts that are available for recombination events. Since multiple transgenic plant events are always produced, one can empirically choose a transgenic event that characteristically has low transgene expression but still protects the host plant from viral infection.

In summary, it appears that growing transgenic virus-resistant plants should not lead to greater recombination and therefore should not result in increased risk of contributing to the evolution of more virulent plant viruses. Although it might be tempting for some naysayers to paint a scary doomsday picture of a GM virus-resistant plant as the Dr. Frankenstein that creates new superviruses, this notion is more suitable for science fiction movies. Nature can and does create novel viruses continually. No new components will be introduced into a transgenic plant that would exacerbate recombination. Transgenic plants, if anything, should delay the evolution of new RNA viruses. While more field data are needed to accurately assess the risks and benefits of growing virus-resistant plants, the data are not needed to make the decision as to whether we should grow GM plants. Virus-resistant plants are environmentally beneficial, if for no other reason, because they would reduce the amount of pesticides sprayed to control vector populations that carry viruses.

In this chapter I mentioned in passing that, like insect-resistance genes, virus-resistance genes in transgenic plants could conceivably be transferred to weedy wild relatives via hybridization and introgression, where they might provide increased fitness. What would happen if we stacked insect-resistance and virus-resistance genes into a single plant, and for the fun of it, added a number of other genes that would confer superpowers to a plant—or at least make it a more competitive plant? What happens if we add fitness-enhancing traits to fitness-enhancing traits? Would we multiply the risks?

10

Superweeds Revisited

Tall Stacks of Transgenes and Waffling Gene Flow

In earlier chapters I alluded to a fictional *deathstar* gene that would be able to confer total resistance to everything for its host plant (figure 10.1). In actuality, the discovery of a single gene that would enable crops to be resistant to all defoliating insects and diseases; protect against heat, frost, freezes, drought, flooding, and toxic metals; and kill weeds that might compete with the crop would be a utopian goal but impossible to accomplish in one fell swoop. Such a *deathstar* transgene could allow the production of diverse crops such as soybeans in the nonirrigated desert and citrus in Iowa. Everyone knows that such a gene does not exist, but finding its closest approximation is the goal—even if it can be discovered in chunks. It should be feasible to build a super plant with numerous transgenes that, when summed, constitute a Deathstar phenotype with total resistance to everything. What are those candidate genes, and what might be the environmental consequences of combining or stacking them in transgenic crops? For example, is it likely that superweeds would be created as the result of gene flow? Or would stacked transgenes be less likely to introgress into weeds than single genes?

Current and Future Transgenes for GM Plants

There are plenty of transgenes known to confer insect resistance. As we have seen, nature is rich in its bank vault of proteins that kill insects. In the species *Bacillus thuringiensis* alone, there are thousands of potential

Figure 10.1. The *deathstar* gene for total resistance ended up having suboptimal specificity. (Drawing by S. Stewart.)

genes that can be cloned, modified, and transformed into plants to protect their hosts against a plethora of defoliating and grain-consuming insects. In the future, we will have GM plants designed to have resistance to nearly every type of damaging insect.

Along with insect resistance, there is significant effort engaged in the discovery of genes for plant disease resistance that could be deployed in GM plants. However, other than virus-resistant transgenic plants using either nucleic acid- or protein-based mechanisms to exploit parts of the virus itself to battle plant viruses, there are no great success stories in transgenic approaches to combat other plant diseases. Given the state of the science, there is still an astounding amount of research to be performed to identify genetic strategies to produce GM plants that are resistant to bacterial, fungal, and nematode pathogens.[1] One fruitful area of research has focused on identifying the molecular mechanisms of how plants naturally protect themselves from diseases. As one might expect, plant defense is a complex and complicated matter. In addition to utilizing protein-based mechanisms, plants use secondary metabolites—

"leftover" chemical compounds that are the product of everyday plant biochemical processes—to fight pathogens. Chemical-based protection of this sort would likely be complicated to engineer into plants because many genes and one or more metabolic pathways could be required. In contrast, protein-based mechanisms are amenable for a transgenic approach, especially if one or a few proteins (hence, genes) could confer resistance. One only has to identify the protein(s) that protect(s) plants from diseases, clone the genes, and then produce transgenic plants. In the hopes of hitting the jackpot, this road map has been followed several times, resulting in a range of success stories; but so far there has been no jackpot, unlike the story of Bt and insect control. Some of the plant genes cloned to date code for antimicrobial proteins, such as phytoalexins, ribosome-inactivating proteins, and pathogenesis-related proteins. Examples of pathogenesis-related proteins are chitinases and glucanases, which disrupt pathogen cell walls.

One fascinating line of research is examining how plants protect themselves from diseases using a natural endogenous process called systemic acquired resistance (SAR), not to be confused with the debilitating human respiratory disease SARS. It has been well established that SAR is the nearest thing plants have to immune systems. Many pathogenesis-related proteins are produced as part of SAR; therefore, a better understanding of SAR will not only enable GM approaches to crop protection, but also lead to improvements in traditional and even in organic methods of crop protection from pathogens. In addition to plant genes and proteins, those from other organisms such as a bacterial chloroperoxidase and human lysozyme have been tested in GM plants. Nematodes, tiny roundworms that predominately affect plant roots, are also considered to be plant diseases. A number of different types of genes have been expressed in GM plants in an attempt to control nematodes. Most of these have their origins in plants and have resulted, at best, in conferring mild resistance to these roundworms. It is clear that obtaining broad pathogen resistance is of great value, but it is biologically complicated. A single gene and protein that protects plants broadly from different classes of diseases (bacterial and fungal diseases and nematodes) may not exist in nature. It seems more likely that multiple transgenes will be required for effective control of these pathogens, a consequence of the complex biology of innate pathogen resistance and the wide variety of pathogens that plants face in their lifetimes.[1] The lofty goal of GM-based disease resistance would be accompanied by a large potential environmental benefit similar to that of GM insect resistance. Chemical solutions to combating plant disease would be replaced by those employing biology and genetics.

Multitudes of researchers are working on a plethora of GM candidate traits for crop improvement. There are candidate genes for heat tolerance, salt tolerance, and tolerance against toxic metals. One big target on which many molecular biology sites are trained is drought tolerance, which would have considerable value to both the farmer and the environment. If plants do not have sufficient water, they either grow suboptimally or die. Farmers sometimes have the option of irrigating drought-susceptible crops, but such an operation is expensive and can lead to deleterious environmental effects, such as the salting-in of soils and creation of water deficiencies in other locations. Researchers have attempted to find genes that, when engineered into crops, would allow the crops to absorb and transport water into the plants as the soil dries out. The best candidates to date have been those coding for osmotic adjustment proteins. Some plants characteristically produce solutes (osmotica) inside their cells, enabling them to draw in water from outside the cells. It would be desirable to engineer plants that osmotically adjust as water becomes scarce. Candidate osmotica such as organic acids and amino acid derivatives have shown small positive effects in the laboratory. The GM plants, when grown in the field, have been disappointingly wilty under droughtlike conditions.

Soybeans are certainly one crop that could benefit from drought tolerance. One team of researchers led by Tommy Carter at North Carolina State University has been using traditional breeding to select for genotypes of soybean that have deeper rooting, which would allow the crop roots to more effectively extract water from the soil. The only problem with this approach is not with the deep-rooted plants, but with the environment. In many parts of the world aluminum is prevalent in soils, and, as stated earlier, aluminum kills plant roots. Unfortunately, even if a deep-rooted soybean were to be produced using conventional means, if it is grown in aluminum-ladened acidic soils, the roots would die, and the resulting plants would not be any more drought tolerant than shallow-rooted plants. There are many transgene candidates that can be used to address aluminum tolerance. One of the most promising is a citrate synthase gene that can be expressed in plant roots, leading to an overproduction of citrate or citric acid.[2] Citrate binds aluminum. When citrate is exuded from roots, it should bind to aluminum and not allow its uptake from plant roots. If the aluminum stays in the soil and is not absorbed by roots, the roots should not be killed by the toxicity of the metal. In addition to citrate synthase genes, there are a number of other plant genes that might be useful for aluminum tolerance.[3] And, in fact, trees and weeds have up to 100 times more aluminum tolerance than do crop plants. It is quite likely that these noncrop genes may be fruitful sources of aluminum-tolerance genes for crop plants, which will lead to corresponding drought tolerance.

Transgene Stacking

What would happen if we put a number of these transgenic traits together—stacked into a transgenic crop? For the sake of argument, let's imagine that we have two different kinds of GM canola plants, each with a unique transgenic trait. One of these plants produces a Bt toxin and is resistant to insects, and the other is drought-tolerant through the use of an aluminum-tolerance gene. We have actually characterized the insect-resistant Bt plant event quite well, and it seems to have no severe biosafety issues. But we know less about the transgenic drought-tolerant plant. It is reasonable to expect that the aluminum- and drought-tolerant canola might be able to occupy sites not possible before; it might actually be able to outcompete several other wild plants that are not as aluminum and drought tolerant. A company that wants to commercialize the drought-tolerant canola would have to thoroughly examine biosafety issues associated with its potentially expanded niche. What if the two kinds of canola plants, insect resistant and drought tolerant, are crossed, so that a new hybrid variety is produced that would contain both traits. This is a case of transgene stacking. Might there be new risks associated with the doubly-stacked GM plant? Or does our knowledge of the ecology of plants with the single transgenes suffice?

Will the risks (and benefits) simply be additive? Stacking is an emerging issue in agricultural biotechnology because combining transgenes is just now being accomplished with an increasing menu of candidate genes. The only transgenic stacks now being marketed by companies are those in cotton and corn: insect resistance stacked with herbicide tolerance or two insect-resistance transgenes stacked together. And there is no reason to believe herbicide tolerance will confer any selective advantage outside of the agricultural field where the herbicide of interest would not be applied. The other pyramid that we've already discussed is the special case of two insect-resistance genes being stacked to give broader protection against different types of herbivores.

Researchers might have two extreme types of opinions with regard to the genesis of transgene stacks. On one hand, there are those optimists who make the case that traits (nontransgenic) are already stacked in plants to a large degree, and there is no special risk in simply adding a few transgenes that do the same thing, only better. Crops already possess some disease tolerance, insect resistance, drought tolerance, and so on. Further, it can be argued that no special provision has ever been made to assess the risk of these tolerant varieties in the past—why should GM plants be treated any differently? On the other hand, there are pessimists who declare that single transgenes are risky enough and that stacking is completely novel. The precautionary principle will be invoked that we simply don't know enough about stacking and that each stack should be thoroughly studied for a long time before the stacked varieties could ever be released. Incidentally, both optimist and pessimist will view themselves as realists. Which side is indeed more reasonable? The answer, I think, lies somewhere in the middle because we find nuggets of truth in the arguments of both sides. Mild "natural" versions of candidate GM traits already exist in plants, and the host crops are generally regarded as safe. For example, some plants are quite freeze tolerant and are not troublesome to the environment. If we were to give freeze-tolerant crops additional freeze tolerance, most people would consider that additional increment ecologically benign because the crop's niche would remain essentially the equivalent and in equilibrium as before. But no citrus trees are freeze tolerant. Conferring freeze tolerance to citrus would result in adding a very profound trait that could alter its ecology and also be economically valuable to citrus growers. Freeze-tolerant grapefruit, in turn, could alter the ecology of other plants if it were to invade northward into colder zones.

Transgene (and trait) stacking is really a question of profundity and degree. We should be able to predict the outcome of releasing single-transgene GM plants before we consider stacks, both to engineered crop and related wild plants to which genes might flow. Before this is discussed further, it will be helpful to explain how transgenic plants are now regulated and how stacked-transgenic plants are regulated in the United States and elsewhere.

Regulatory Lesson from the School of Redundancy School

As it stands now, the government of each country is responsible for independently regulating transgenic plants. If there is much cooperation among nations (with possible exception of EU nations), it is imperceptible to me. I briefly outline the responsibilities of each U.S. regulatory agency and also briefly discuss how Canada approaches the regulation of transgenic plants differently.

In the United States, three agencies are responsible for oversight of GM plants: the Food and Drug Administration (FDA), the EPA, and the USDA. There are subagencies within each agency that are responsible for different aspects of regulation. I won't go into those, other than to denote that the USDA subagency is called the Animal and Plant Health Inspection Service (APHIS). A fourth agency, the National Institutes of Health (NIH), has guidelines that are followed in all laboratories where recombinant DNA is manipulated. The FDA is responsible for food safety. It is responsible for determining if any new food or food additive is safe for consumption and therefore for ensuring that it is not toxic or allergenic. We are primarily concerned with ecological and not food biosafety in this book so I will not discuss further the FDA's role in GM plant regulation.

The primary role of the EPA in the regulation of transgenic plants is to provide federal oversight whenever GM plants produce their own "plant-manufactured pesticides." The most visible role the EPA has played is in the deregulation and subsequent oversight of GM Bt plants. The EPA sees Bt crops with the lens that reduces them to a vehicle that produces a pesticide. Therefore, since the EPA is responsible for registering pesticides for different crops, it follows that their expertise would convey to regulating Bt crops. The following illustrates how registration works. Let's say you have Japanese beetles damaging your rosebushes. You would like to spray an insecticide on them to kill the exotic pests. You will go to the store and read the pesticide labels on insecticides approved by the EPA (registered) for the use of controlling Japanese beetles on roses; the label will describe the safety issues of the pesticide and how it should be used. The EPA collects data from pesticide companies to assure consumer and environmental safety from using the insecticide on roses. When sufficient data convince the EPA that the pesticide is indeed safe, the company can now sell it for its specific use.

Likewise, the EPA also registers pesticidal plants just as if they were chemical pesticides. The EPA has a wealth of experience in toxicology and approving or disapproving pesticides to be used on crops and other plants. On one hand, it makes a great deal of common sense to take advantage of EPA's experience in requiring data on nontarget effects and other effects that could negatively impact the environment. On the other hand, scientists have complained that EPA has no business with oversight responsibility of Bt crops. They argue that Bt crops that are just that—simply crops—and that there is no pesticide application in the picture. But it really doesn't matter what scientists say on whether or not the EPA should have oversight. At this time, the EPA has statutory oversight for plants that "produce pesticidal compounds." Once the EPA is involved (i.e., if an insecticidal Bt gene is engineered into a crop), then it also requires the company seeking registration and deregulation (commercialization) to provide data on gene flow and wild relatives, toxicology, and nontarget effects, as well as other data for its risk assessments. Periodically, GM plants come up for re-registration. I have sat on the EPA Science Advisory Panel and have seen how this agency operates first hand. In my opinion, I think it is downright silly to regulate a plant as a pesticide and that the EPA should not have redundant regulatory oversight with the USDA in this manner. However, given the constraint of EPA's statutory oversight in this regard, I have no qualms about how the EPA does its job in the practice of regulation. Their scientific advisory panels consist of scientists with diverse opinions, backgrounds, and areas of expertise. Some activists have charged that the EPA kowtows to the agricultural industry. I have seen no evidence to support their assertion, quite to the contrary, in fact.

If a plant is not pesticidal, then the only regulatory agency in the United States responsible for ecological biosafety issues is the USDA. Its purview of oversight includes not only deregulation issues, but also the interstate shipment of transgenic seeds and the growth of experimental GM plants in the field; it is responsible for all things ecological for all GM plants: side effects, gene flow, unintended consequences, and so on. So, for Bt plants, there is some redundancy of oversight between the USDA and the EPA once a GM plant is nearing commercialization. At first glance, this arrangement seems more complicated than it should be. But the fact of the matter is that some duplication of regulatory oversight is a result of historical regulatory perspectives, the expertise now possessed by regulators, and also a result of the relative immaturity of the regulatory processes. Although some people might charge otherwise, it is clear that the current oversight policies are extremely conservative in the United States. Redundancies in oversight have resulted in stricter regulations and more regulatory intervention than if the processes had been streamlined and properly carved up at the outset. Just as the new Department of Homeland Security does not yet really "own" antiterrorism efforts (these duties are distributed and, in many cases, duplicated across many agencies), it is unlikely that there will be any one office in Washington in charge of regulating and deregulating GM plants.

Current Regulatory Climate

The various U.S. regulatory agencies have done an admirable job of regulating GM plants in the way they've been charged to do so. There have been very few mistakes and course corrections. But because the United States is the global leader in GM plant technology, the global community expects an equally advanced regulatory structure and performance. The United States would have the most to lose if some major regulatory faux pas, akin to Britain's outbreak of mad cow disease, were to occur from a GM plant ecological disaster. I'm surprised more of the world does not replicate the American system of regulation. The Europeans have taken a route that is more precautionary and political in nature, and the precautionary principle is invoked effortlessly as *the* philosophical directive in regulating GM plants by the European Union. One problem with the term "precautionary principle" is that it has several different definitions and interpretations depending on circumstance and whimsy. I do not want to dwell too much on the precautionary principle, other than to say it is usually invoked to say, "we don't have enough information to prove that GM plants are absolutely safe (or risk free, or economically favorable, etc.), so we should not grow or eat them. Let's not be so hasty in adopting this new technology." Entertainment of such arguments from people hiding behind the precautionary principle in total avoidance of technology has realized two outcomes in Europe: (1) Not only has the implementation of progressive regulation stalled, but (2) science and technology has also been hindered as seemingly endless debates have ensued. It is irresponsible not to weigh benefits against risks in practicing and adopting technology. The consequences of exclusively focusing on risks relegate one to abandoning benefits of an nonadopted technology. A precautionary principle-based regulatory structure has not proven favorable to innovation, science and technology, research and development, or science-based regulation. And I would add that the opportunity costs of slavery to the precautionary principle rules out opportunity for technology-based environmental progress as well. Nations are moving toward implementing science-based regulation at various paces that seem commensurate with the maturity of their science and biotechnology in general. And the maturity in the science and regulation of GM plants is in its adolescence. Rules, transgenes, and crops are ever-changing, and the regulations seem to be in flux, too.

I would characterize the current regulatory climate in the United States as one of cautious confidence. The country is growing a record number of GM plants (more than 7 trillion in 2003), with no documented environmental or human health harm, but there seems to be increased political pressure for sensitivity and caution. In one example, the White House Office of Science and Technology Policy suggested in 2002 that there should be increased regulatory oversight by the FDA, EPA, and USDA to minimize trace amounts of transgenic proteins stemming from GM plants. This move is likely being driven from the present and future production of pharmaceutically active compounds in plants, as nearly everyone is convinced that seeds from the "pharm farm" should not be mixed with regular seeds. (How to ensure that commodity mixing does not occur is the trick. I think pharm farm plants should simply be grown in greenhouses, or at least in areas where commodities are not grown, but that's another story). In another example, the National Academy of Sciences was commissioned to study the USDA as a regulatory body of GM plants and give recommendations for changes. The report was authored by a committee chaired by Fred Gould, who is known for his outstanding work on Bt-resistance management.[4] The committee was composed mostly of ecologists, but it included a few molecular biologists and even a philosopher. As with most committee reports, the results were somewhat predictable on the basis of committee composition. The recommendations focused on the need for more research in agriculture, post-commercializing monitoring of GM plants, closing of some holes in the USDA's principles and practices of regulation, and more regulatory transparency by the agency. The gist of the report was that the USDA was doing a good job, but there were several recommendations for improvements. In addition, the report also stated that the USDA should be sponsoring more biotechnology risk research. A great majority of the committees' suggestions would require additional resources to implement, resources that will almost certainly not be readily forthcoming in light of economic downturns and the overall depauperate congressional funding appropriated to the USDA and agriculture. Still, it is notable that in Gould's preface in the report, he states "we found that the current standards used by the federal government to assure environmental safety of transgenic plants were higher than the standards used in assuring safety of other agricultural practices and technologies. After much discussion of this finding we did not conclude that the standards for transgenics were too high."[4] Indeed, the committee recommended additional strengthening of the USDA's standards and practices in regulation; once again, a result predicted by a committee consisting of a significant number of members with anti-biotechnology sentiments. In the past I have waffled between the opinion that transgenic plants are overregulated in comparison to conventionally produced plants and the inevitability that novel transgenes in certain crops warrant certain special regulatory scrutiny. I am becoming increasingly convinced that GM crops are under too much regulatory oversight because they are simply transgenic (indeed this opinion is gaining support among the scientific community). The status quo ignores the fact that there is nothing dangerous or risky per se about being transgenic. That said, I agree on one finding by the committee on transgene stacks. They recommended that the USDA change the way it regulates transgenic events.

The Main Event

Currently the USDA and the EPA regulate, deregulate, and register transgenic events. To review, a transgenic "event" is the insertion of a transgene in the genome in a plant cell that subsequently leads to a GM plant line with the transgene in a specific and characterizable chromosomal location. The GM plant that comes off tissue culture is dubbed an event. It is defined by its transgenic construct (promoter, gene, and terminator) in a specific spot in a chromosome that is produced by an engineering event in time. The event, by definition, also encompasses the progeny line—even if the descendants end up as different species (think about transgenes flowing from canola to a wild relative, as described in chapter 4). So, a single event of transgenic canola can be crossed, then repeatedly backcrossed with field mustard, and the hybrid line is still defined as the identical GM plant event, even though the genetic background has totally changed. Therefore, under current USDA rules, if a transgenic event for a Bt gene is deregulated, and a transgenic event for herbicide tolerance is deregulated, one could cross these two plants and have a new and novel plant that would be, by definition, deregulated with no new scrutiny from the agency.

We can imagine all kinds of transgene stacks that might behave differently in a host (especially if the host is a totally different species!) compared with single genes. For example, heat-tolerance genes could allow a crop that currently has little insect pressure to be grown in more tropical, warmer climates where insects abound. A Bt gene in these tropical locations would have a greater fitness effect on the host than in cooler places, since insect pressure would generally be greater in the tropics. Interactions and synergies would abound in several scenarios. I predict that the transgenic crops grown in 2020 will have multiple genes for insect resistance, disease resistance, herbicide tolerance, and drought tolerance. And I am probably shortsighted. If the United States (and USDA) is continuing down its "transgenics are special" paradigm, the unit of regulatory concern should be adjusted to take into account stacks and novel phenotypes. The Canadian regulatory system, indeed, focuses on novel phenotypes and traits. Canada does not see genetic modification as worthy of regulation in and of itself; for example, it regulates herbicide-tolerant phenotypes produced by GM, by random mutagenesis, and from wide conventional crosses equally because the trait is novel. But in the United States, the transgenic event is the unit regulated by the USDA for shipping, releases into the environment, and petitioning for deregulation.

Useful regulation should be adjusted to reflect the genetic and ecological ramifications of how *traits* can interact with one another, or even how two very strong insect resistance transgenes could interact with a plant's ecology. For example, the stacking of multiple transgenes for insect resistance could transform a plant into an insect-killing machine with broad host range toxicity (back to the *deathstar* picture). Under current regulations, an event that kills insect A could be crossed with an event that kills insect B, and the progeny could be crossed with an event that kills insect C, and so forth, until the new plant kills insects A through Z. The companies that produce transgenic plants understand this fact, and so do many other countries that regulate the final product and not just individual components of the product. Therefore, even though the USDA's regulations need tweaking, its current status probably does not affect today's transgenic crops. But we are looking to the future of inevitably stacking transgenes in commercially available plants. And if we are going to regulate transgenic plants, the regulations should be as helpful and effective for the plants of the future as they are for today's plants. Although broad-range insect-killing capability may affect nontarget species, it is unlikely that the transgene stacks will be easily moved from crop to wild relative. So, transgene stacking could endow plants with additive risks of one type (nontarget effects) but less-then-additive effects of another type (transgene movement and persistence into wild relatives).

Back to Genetics 101: Transgene Stacks = *deathstar*?

There are good reasons why companies are pursuing transgenic plants with single and stacked-gene traits. They produce GM plants because there are customers who find value in genetically improved crops, foods, and environmental applications of biotechnology. The traits (genes) are chosen because they add value to the host (e.g., crop) and are easy to work with, and for numerous other reasons beyond the scope of this book. How the genes are actually deployed in the GM plant genome is open for discussion. Different deployment (placement) strategies have a wide range of advantages and disadvantages. There are several ways to put stacks of transgenes into plants, and we will take a closer look at two extreme scenarios.

Let's imagine we have determined the identity of 50 novel transgenes that we can engineer into a plant that would give us the perfect plant; a plant with total resistance to everything. It would need no pesticide, would defend itself against weeds, could not be frozen, and so on. In one scenario, we could place novel genes 1–50 all on a single plasmid (currently not technologically possible because there is an upper limit to the amount of the DNA that can be cloned onto a single vector and transferred into the plant genome; but we won't stress that fact too much because that technical hurdle will most certainly be eventually jumped, perhaps with the production of artificial plant chromosomes), which would result in all genes being located in one place on one chromosome in the host plant: one event. Alternatively, the 50 genes could be engineered onto separate plasmids and engineered in separately, so we would have 50 different transgenic plant events. In the latter scenario, the new transgenes will be located randomly throughout the genome on different chromosomes (figure 10.2).

Of course, with a single transgene (and a selectable marker such as kanamycin selection), we have no practical choice but to put the genes on one plasmid with an insertion into a single chromosome. Having all the transgenes on one plasmid (and one chromosome) makes the situation advantageous for plant breeders who can move the suite of genes together into elite varieties. One simply makes crosses and selects on the transgene over and over through multiple crosses while also selecting for all the advantageous traits in the elite varieties. As mentioned earlier, there

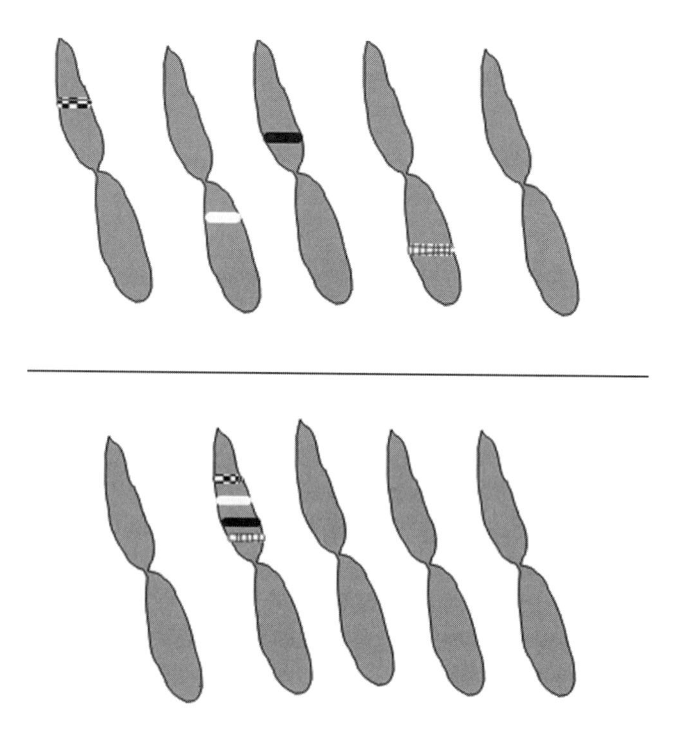

Figure 10.2. Two different engineering scenarios with a few transgenes. In the top scenario, transgenes are introduced individually and are located randomly throughout the plant genome. In the bottom scenario, transgenes are linked and introduced together. They are in one location in the genome—on a single chromosome. (Figure by Reginald Millwood.)

are more than 1000 different soybean varieties with the Roundup Ready gene for herbicide tolerance. But putting all 50 genes on one plasmid to begin with is difficult, to say the least. The bench scientists would wind up doing lots of engineering of DNA constructs to be singly transformed into plants. I suppose if one wanted to revise the GM plant and add additional genes afterward, the bench scientists would have to add the newcomers to their vector and make new transgenic plants. Then the breeding must be performed over and over, each time a new transgenic event is made that is moving us ever closer to *deathstar*. So, unless all the genes are identified at once and cloned at once, this alternative will be expensive. And it will be easier to move the final *deathstar* conglomerate into different varieties compared with the second alternative. And it is intuitive that having a suite of genes that add up to a composite *deathstar* gene for total resistance to everything may have negative environmental effects.

The second alternative results in a random distribution of transgenes throughout the genome among chromosomes—50 different events followed by much plant breeding. In this case, the genetic engineering is easier because genes can be simply added through crossing into one crop variety. If we have just a few elite varieties, they might be engineered directly (a lofty goal, but one that has seen much progress), then we have to breed over and over again to cross in all the genes. The potential problem here is the selection of multiple genes on different chromosomes after every cross. But it is important to realize that even if there were multiple genes stacked into a single plant (a weed even), that the power of *deathstar*'s potential negative environmental effects would be diminished. It all has to do with the genetics.

It's the Genetics, Stupid

When the Roundup Ready gene was bred into the numerous soybean varieties, crop scientists discovered that the yield of all the varieties containing the new gene was less than the parent varieties without the transgene. These results were found in absence of weed pressure and Roundup applications. Of course, farmers live in a world with weeds, so in the real world the Roundup Ready varieties outperformed traditional varieties. But what was causing the "yield drag" in the herbicide tolerant varieties? Was it because of the transgene itself and the protein it coded? Did the new gene carry with it an energetic or metabolic cost or drain? The evidence suggested that, no, this was not the explanation of the observed yield drag. The most likely cause could be traced back to the original transformation event that was created in the 1980s. Whenever a transgene goes into the genome, it must land in a chromosome somewhere, and its landing spot might even disrupt a resident gene. *Agrobacterium* tends to insert genes in transcriptionally active areas, so there is a significant chance of some type of genomic disruption. The goal of the genetic engineer is to produce so many transgenic events that he ends up with one excellent line. But in the 1980s, transformation procedures and efficiencies were not as good as they are today, and there were corporate pressures to get the GM soybean varieties to market. Remember, too, that events are the entities that are deregulated by the USDA, not just transgene and crop combinations. So Monsanto had to dance with the date it chose pretty early on. To change events to a better yielding one would require jumping through significant regulatory hoops that would be very expensive. When breeders moved the Roundup Ready transgene from the original plant into other varieties, the yield drag due to the genetics around the transgene came along for the ride. With apologies to John Donne, no gene is an island.

Superweeds Revisited

The following discussion is largely theoretical but is based on our present knowledge of the science and technology in plant transformation and biosafety. Although it is only theory, it will be very important as we think about deploying transgene stacks in plants and perform experiments that elucidate the consequences of transgenes residing in weeds and wild relatives of crops. The following discussion is also important to engineered plants with single transgenes. As noted in chapter 4, my research group and collaborators have performed experiments that examine the effects of the expression of a Bt gene for insect resistance in canola (*Brassica napus*) and have studied the hybridization and introgression of that gene and a marker gene in wild relatives, especially field mustard (*Brassica rapa*).[5,6] When we move the transgene into the weed, let's say after three to four crosses to the weedy species, we determined that the new transgenic weed has the same number of chromosomes as the non-transgenic weedy parent. But what we have come to realize is that what we are calling transgenic field mustard also contains at least part of the chromosome that carries the transgene, along with, perhaps, other chromosomes (or parts thereof) that were originally from the crop. So, what we believed to be an introgressed transgenic weed, is really more "croppy" in its genetics than the non-transgenic weed.[7] Many domestication genes were, no doubt, coming along for the ride with the linked transgene. This fact can affect the interpretation of experiments, and, I believe, be important in biosafety considerations.

When we performed artificial greenhouse experiments, we competed Bt *B. rapa* (the crop backcrossed three generations with field mustard) with soybean and non-Bt *B. rapa* with soybean, with or without insect treatments. Present in some plots were defoliating insects that would eat non-Bt weeds but would not eat soybean or Bt weeds. We found that the Bt weeds outcompeted the non-Bt weeds when defoliating insects were present. That is, the soybean grew much better when in competition with non-Bt weeds compared with transgenic weeds. The initial conclusion was that the Bt gene made field mustard a weedier weed when defoliating insects were present compared with the non-transgenic *B. rapa*. We performed this preliminary experiment as a first pass to test our experimental approach. We wanted to determine if a specific transgene enhanced weediness in transgenic weeds by measuring crop performance using competition experiments. If the transgenic weeds were weedier than non-transgenic weeds, the crop yield should be reduced when competed against transgenic weeds compared with regular weeds. We knew the greenhouse experiment would not realistically predict the performance in the field. First of all, for a host of reasons, greenhouse experiments never fully do. But, second, we added too many insects (so they ate an unrealistic amount of weeds). And, finally, field mustard is not a weed on soybeans. But the preliminary conclusions we drew were that we did make a weedier weed by adding the Bt gene to field mustard when *Brassica*-eating insects were present and that our experimental approach was sound. After two recent realistic field experiments, we are beginning to realize how genetics complicates erstwhile predicted ecological effects.

The field experiments consisted of taking three different populations of Bt-backcrossed weed hybrids and non-Bt weeds and competing them against wheat.[8] The Bt weeds were populations of field mustard (*B. rapa*) × canola hybrids backcrossed two generations to *B. rapa*, then bulked in two more intratype crosses. We selected one population to contain the transgene and selected against the transgene in the other population.

Like the greenhouse experiments, we assessed the weediness potential of our new transgenic weeds by measuring crop performance relative to when wheat was competed with regular non-transgenic weedy field mustard. We found the opposite result compared with that from the soybean experiment. The Bt weed lines actually competed less well than wildtype weeds against wheat. But in this case, insects were at ambient levels, not at the supercharged numbers in our greenhouse experiments. Once we began to analyze the data, we realized we had a bunch of crop genes in our transgenic weeds, and we could quantify the number of genes in the crosses coming specifically from the canola parent using molecular markers (15–29%). Eventually, crop genes would be purged from the backcross hybrid weeds, but it would take a long time, many generations, and strong selection for that to occur—not just three generations. The real question is, could the backcrossing continue in light of superior competition? Yes, it is possible that transgene introgression could happen in *B. rapa*, but would it under realistic field conditions? It is important to realize that weeds are highly adapted to compete with crops and grow as weeds—that's why they're weeds to begin with! The suite of genes they possess make them successful weeds. If we introduce a transgene that might help them survive attacks by insects, drought, and so on but transferred nonhelpful, nonweedy genes (crop genes) as tag alongs, it might make the resulting weed less competitive than its parent. In essence, our experiments were more about partial domestication of weeds than about the creation of superweeds.

These were the transgenic weeds one writer cleverly likened more to the mild-mannered Clark Kent than to Superman.[9] So our experiments are now going back to the proverbial drawing board for some weed breeding and selection to see if we can better replicate, experimentally, the types of selection that mother nature might perform once a transgenic weed develops on its own if Bt canola were to be commercialized. It is clear that the genetic load of crop genes will be an important force in transgene hybridization and introgression—whether transgenes can progress through these genetic bottlenecks or not.[10] The species boundaries from crop to weed in this and many cases are significant. There are chromosome and genomic properties that keep most species intact and distinguishable.[10] While interspecific gene flow does occur, we should not expect anything special from transgenes. The genomic surroundings around the transgene, taken together, might prove to be much more important than the transgene itself.

No gene is an island. The effects and selection of transgenes will be moving and asserting effects against the backdrop of other plant-gene linkage effects. And it has been recently shown that in many crops, there are regions of chromosomes that are introgression dead zones. Statistically speaking, plant genes in these dead zones are introgressed into compatible wild plants at a much lower rate than would be predicted by chance.[10] Experimentally and theoretically, we can estimate the effects a single transgene will have on its host, but these effects will be moderated by inherent genomic interactions that occur in non-transgenic and transgenic plants alike. It is important to realize that these larger genomic effects have taken place for eons, and there is nothing new under the sun. There is no reason to believe transgenes will behave differently than plant genes.

When we then extrapolate to gene stacks with the suite of genes distributed randomly throughout the genome of a plant or weed, there is a great chance that early hybridization and introgression events between weeds and crops could actually produce less-fit crop–weed hybrids that will not enjoy the full benefit the multiple transgenes could potentially confer if they were located together on one particular chromosome. In addition, the stacks will almost certainly benefit crops more than weed plants because they are often domestication related. In other words, multiply stacked transgenes that are randomly distributed would not additively constitute a *deathstar* gene that could make a weed a superweed. Transgenes would become separated from one another. Genetic modification, it seems, is much better for crop improvement than weed improvement because of the genetic load of crop genes that must tag along with transgenes as they make their way into weed genomes. The "no gene is an island" idea is not new. This effect, called linkage disequilibrium, has been long identified but has been largely ignored by scientists working in the environmental biosafety of GM plants.

To conclude, we now have more information to do better experiments and to ameliorate worries about the creation of superweeds. We still need to do the experiments on a case-by-case basis, but it is becoming increasingly clear that to perform the most relevant experiments, we need to understand the genetics of transgenesis, hybridization, and introgression as much as or more than we need to understand the ecology of gene flow and competition. In addition, if natural barriers to introgression are coupled with engineering flanking transgenes with known domestication genes, and these are coupled with GURT, the effective hurdle to introgression could be impossible to jump.[10]

The next chapter surveys the world of environmentalism and environmentalists. We will look at the most recent environmental movement, and why environmental activists are nearly unanimously opposed to genetic engineering in principle and practice, and some examples of what it means to be green.

11

Green and Greener

Environmentalism, Agriculture, and GM Plants

GM plants on whole have proven to have a great many environmental benefits and few risks. While no technology is without risks, in most cases the actual risks of GM plants are quite comparable to those of growing conventional varieties used in everyday agriculture. But there are many people who universally condemn transgenic plants as the products of dangerous technology. They argue with those who focus mostly on benefits GM plants offer. Such broad-brush judgments pigeonhole science and polarize discussion. The intricacies of biology make necessary risk–benefit analysis for transgene–crop combinations on a case-by-case basis, and those in the middle of the risk–benefit analysis accept that science and technology have value in improving human lives and the environment. Although there is plenty of room in the middle ground between opposed extremes, environmental benefits of GM plants have gotten short shrift, especially from environmental activist organizations.

This book has been mostly devoted to scrutinizing risks as described by scientific research. En route, it has been impossible to ignore the importance and magnitude of the environmental benefits transgenic plants can provide. For today's crops, environmental improvements range from decreased use of traditional pesticides (chemicals having known side effects) to higher productivity (requiring less land to produce the same amount of food) to decreased tillage for crops tolerant to herbicide applications. These benefits can be quantified and have established economic value. For example, in 2000, 28.16 million acres of Bt crops were grown worldwide.[1] Cotton, the number one crop for insecticide usage, has 71%

of its worldwide acreage in developing countries. Seventy-five percent of the global cotton crop is treated with chemical insecticide to control damaging caterpillars. Therefore, not only is Bt cotton for caterpillar control beneficial for decreasing insecticide sprays, but it could also aid in the developing world's independence from chemical inputs. This same report stated that the USDA National Agricultural Statistics Service had determined that, in the United States alone, there was a reduction in insecticide use of more than 2,000,000 lbs in 1998 and 2,700,000 lbs in 1999 as the result of growing Bt-transgenic crops. This drastic decrease in the use of insecticides is the fulfillment of Rachel Carson's dream, in which biological solutions would replace chemical solutions for pest control. These are real environmental benefits that outweigh current risks of today's commercialized transgenic crops.

Despite the growing canon of data showing high biosafety of GM crops, there are still those who doubt that growing GM plants is good for agriculture and the environment. Most objectors have misgivings about the ecological biosafety of GM plants. But the science is clear that biotechnology is not inherently hazardous. Like most science and technologies, it is the implementation of biotechnology where practical safety intersects with utility. In contrast with reality, in which commercialized transgenic plants have extremely low environmental risks, and for reasons that transcend science, many activists continue to promulgate the message that biotechnology is an unavoidably dangerous science.

It seems to me that if GM technology is not dangerous and that it could actually benefit the environment, it works against the greater cause of environmentalism if activists oppose biotechnology wholesale. For that reason I've frankly been somewhat confused about environmentalists' zealous opposition of GM plants. I must admit that I came into writing this chapter "green" with ignorance about many of the motivations of environmental groups with regard to GM plants. As a scientist there seemed to me to be a disconnect between what we know from science gathered by biotechnological and ecological risk research and environmental rhetoric. It seemed to me that rationally and scientifically, environmentalists should generally be in favor of any incremental help to the environment, no matter what package it came in, even if it were a biotechnological augmentation. But why aren't they? A brief history lesson should shed some light on this issue. To better understand why environmental activists are the chief opponents of GM plants, it will be helpful to review the history, philosophy, and underpinnings of the modern environmental movement.

Beginnings and Underpinnings

Rachel Carson and Paul Ehrlich head the list of intellectual and scientific catalysts in the 1960s and 1970s who formed the roots of today's environmental movement, which is generally antiagriculture. But why be antiagriculture? From Rachel Carson's point of view, chemical pest control and agriculture went hand-in-glove, and thus farmers became the scapegoat for environmental travesties. Farmers were the end users of chemicals that had environmental side effects. But from the farmer's perspective and experiences, doomsday predictions did not ring true. Dire environmental scenarios did not seem plausible to an agronomic society (and many would say it was in a postagronomic/industrial phase) in which most people had recent family memories of life on the farm. Farmers never perceived farms to be polluted sites or point sources for environmental contaminants; conventional wisdom of agriculturalists held that farming was a close-to-the-ground environmental experience and that farmers were mostly environmentalists in practice. The farmer began each morning with a tromp across the cool, dewy field to tend animals and perform other chores. Agroecosystems had to be sustainably managed because the farmer's livelihood relied on having healthy soil, clean water, and fit plants and animals. Farming, indeed, is a biologically rich and sensual existence, in contrast with the urban and suburban lifestyles of those who were at the forefront of the environmental movement. For the urban dweller, the environment is perceived to be an external entity—out there someplace—but for the farmer, the environment is in his backyard. Indeed, the early environmental intellectuals did not go far in winning the hearts of middle America when inflammatory terms such as "agricultural propagandist" were brandished, accompanied by copious amounts of condescension.[2] Environmentalist ideals were slow to immediately affect those outside their tight circles in the 1960s. Ehrlich's ultimate societal and environmental solutions, which primarily focused on the perceived problem of overpopulation, bordered on totalitarianism that one would not expect to be well received in any capitalist and socially conservative society, and they weren't. Most Americans did not embrace ideas they feared would revolutionize their country into one with forced population control, coerced abortions, sterilizations, and euthanasia. In the midst of the Cold War, these were the well-known characteristics of many communist regimes.

In spite of the initial sketchy reception from middle America, there were merits to the arguments made by Erhlich and like-minded idealists that demanded immediate fundamental changes in environmental policy and laws. People had viewed environmental disasters of tremendous proportions broadcasted on the nightly television news. These catastrophes included horrific oil spills and fish kills that terrified many people to the same degree as the graphic bloodletting that was occurring a world away in Vietnam. But many of the environmental disasters were occurring in the United States and in places close to home. Environmentalism not so slowly worked its way into Americans' hearts, ideas, and schools. The environmental movement was energized by the intense idealism and energy of rebellious youth that was loosely connected with the hippie movement and West Coast ivory towers. Environmentalism was only one of the components of liberal politics that demanded quick change in American mores. The underpinnings of the environmental movement, however, rested in real problems that had come not only from neglect, but also from technology and affluence itself. As mentioned before, these included agricultural technologies.

Birth of a Movement

The environmental movement was certainly not entirely new in the 1960s. John Muir founded the Sierra Club in 1892. By the middle to late twentieth century, the environmental movement began to make policy changes in government. Ehrlich called for the formation of the Department of Population and Environment in 1968.[2] The EPA was created in 1969 with the passing of the National Environmental Policy Act. DDT was banned in the United States in 1972 (although it is still manufactured in the United States for export), and, indeed, air and water quality improved thereafter. The new environmentalists called for and caused a cascade of rapid improvements in environmental quality. Certainly, doomsday scenarios of mass famines, political unrest, and economic collapse mobilized swift action, and on April 22, 1970, the first Earth Day was celebrated to consummate the early success of the environmental movement. But more than that, rapid-fire successes fed a call to even more action and activism.

Organized environmental activism was largely born in the early 1970s. There are several such organizations in existence today that were largely created in the 1970s ranging from Friends of the Earth and the Earth Liberation Front to the epitome of green groups, Greenpeace.

Greenpeace

In the words of Patrick Moore, cofounder of Greenpeace in 1971, "I became a born-again ecologist, and in the late 1960s, was soon transformed into a radical environmental activist."[3] Originally, Greenpeacers came together to protest nuclear weapons testing. Indeed, they were successful in helping ban nuclear weapons trials. Moore explains,

> Flushed with victory and knowing we could bring about change by getting up and doing something, we were welcomed into the longhouse of the Kwakiutl Nation at Alert Bay near the north end of Vancouver Island. We were made brothers of the tribe because they believed in what we were doing. This began the tradition of the Warriors of the Rainbow, after a Cree legend that predicted one day when the skies are black and the birds fall dead to the ground and the rivers are poisoned, people of all races, colors and creeds will join together to form the Warriors of the Rainbow to save the Earth from environmental destruction.[3]

They named their sailing vessel after this ancient prediction. During the 15 years of Moore's involvement in Greenpeace, it became the world's largest environmental activist organization.

From nuclear testing protests, Greenpeace took on Soviet whalers with showy and risky moves by floating its craft between harpooners and their targeted cetaceans. It went to the front lines to interfere with hunters who clubbed baby seals for pelts. Among other things, it protested the dumping of medium- to low-level nuclear wastes into the ocean. These were all high-profile targets that led to plentiful opportunities for effective public relations for the movement. Greenpeace and other environmentalists were clearly the good guys, and industrialists, hunters, and the military were the bad guys. Moore states that by the mid-1980s, Greenpeace had revenues of more than $100 million per year, with offices in 21 countries and more than 100 campaigns around the world. It was protesting "toxic waste, acid rain, uranium mining and drift net fishing as well as the original issues."[3] Greenpeace was a great success. As a teenager, I cheered the *Spirit of the Rainbow* during the ship's high-profile adventures because it seemed Greenpeace's objectives were righteous and justified. And with each adventure, the environment as a whole had won, and the world became a greener and happier place.

Declinations

Moore is now a Greenpeace expatriate. Why did he finally abandon the organization? For the same reasons that Greenpeace and other environmental groups, I think, protest GM plants: extremism for extremism's sake. Some people have made the argument that protesting biotechnology is an effective tact for fundraising. In the words of Moore, Greenpeace "rejected consensus politics and sustainable development in favor of continued confrontation and ever-increasing extremism. They ushered in an era of zero tolerance and left-wing politics" to the environmental movement.[3] This buildup of power all happened before the genesis of agricultural biotechnology. Because GM plants were mainly championed by large agrichemical companies that were already a target for activism, it was logical that environmentalists would oppose this new and revolutionary product. The same companies that assured the public that chemical pesticides were wholly safe also touted biotechnology as environmentally friendly—why should they be believed? In addition, activists (and others) loathed the industrialization of agriculture dominated by a few multinational companies. But even though many farmers don't relish being driven by mega-agricultural industries, they see no reason to throw the baby out with the bathwater. And while environmentalist activists have proffered many admirable causes, farmers are befuddled by environmental extremism that could realistically make farming impossible.

Environmental Extremism

Environmental extremism is not the exclusive property of Greenpeace. As we'll see later, some environmental groups are far more extreme than Greenpeace. As a backdrop to understanding why extreme environmentalists are dead-set against GM plants and even agriculture in general, it will be helpful to have a better understanding of the philosophy that underpins the extremism woven throughout segments of all major environmental groups. Moore has provided a concise list of the characteristics of environmental extremism. The list and quotes are from this Greenpeace expatriate, but I have amplified how these characteristics impact agricultural biotechnology.[3]

1. Anti-humanism. "Humans are characterized as a cancer on the Earth." This anti-humanism devaluates many human activi-

ties. Especially evil would be those activities that exist for the sole purpose of increasing human populations and perpetuating health. The chief of these activities are agriculture and medicine, the two fields in which biotechnology has had its greatest impact.

2. Antitechnology and antiscience. This point is simply a corollary of the first characteristic. Science and technology, to some extent, are human centered. It is not surprising that extremists never let science speak for itself in an objective fashion. Instead, there is a tendency to cherry-pick scientific results to justify a doctrine or preconceived dogma. Extremists are renowned for taking science out of context to interpret results in their point of view. Moore states, "Unfounded opinion is accepted over demonstrated fact."
3. Antitrade and antibusiness. Corporations are considered to be corrupt entities driven by greed. Capitalism is continually pitted against the betterment of the earth. The two are not compatible in the eyes of the extremist. Is it any wonder that Monsanto, the American-based multinational vanguard of biotechnology, has been an ever-popular target of environmentalists?
4. Anticivilization. Moore states, "In the final analysis, eco-extremists project a naïve vision of returning to the supposedly utopian existence." He says further that they "conveniently forget that in the old days people lived to an average age of 35, and there were no dentists. In their Brave New World there will be no more chemicals, no more airplanes, and certainly no more polyester suits." Their ideal for agriculture totally rests on the shoulders of organic production, complete with the constraints peculiar to its practice. However, organic agriculture cannot feasibly supply all our calories. There is not enough land or enough available nitrogen to support more than a couple billion people on earth, and then, even it were possible, I think we'd all have to be farmers.

Are most environmentalists extremists? I don't think so. Bona fide environmentalists universally value clean air, water, and conservation of wild lands, as well as other properties of nature that lead to stable ecosystems and a high quality of life.

Where Scientists and Science Fit into Environmentalism and GM Plants

Of course, many components of environmentalism transcend science. There are plenty of religion and sociology experts who are better qualified than I am to explain why deep ecology and paganism might cause groups to not be naturally inclined to embrace the virtues of genetic engineering. For the next section, I would like to turn away from environmentalism per se and turn to science and scientists and their impact on the environmental rhetoric. Within science, there are myriad voices impacting exactly what biotechnology research is performed and then how it is interpreted. The field of biotechnology is a great deal larger than the nuts and bolts in the toolbox of biotechnology. Ecologists, agriculturalists, and entomologists, among others, have crucial voices in the science of agricultural biotechnology. I want to take a brief look at how different types of scientists typically see the GM controversy and how they might interact with environmentalists (who are typically not scientists).

There are three scientific/quasi-scientific viewpoints on the GM plant debate spectrum. On one side, there are industry scientists (chapter 13), who are generally pro-GM; their livelihoods depend on it, but they wouldn't have taken a job with an agriculture company if they were not sold on the benefits of agricultural technology. Although most people may assume they must be scientifically biased, in my experience, industry employs some truly exceptional scientists whom one can engage in purely scientific discussions. As a rule, they approach science objectively. On the complete opposite end of the spectrum lie environmental groups, which range, as we'll see below, from quasi-rational/quasi-scientific to totally nonscientific in their approach. They are the antimatter to industry's matter, but, in contrast, they generally have few to no people on their staffs with scientific credentials and expertise. Industry scientists perform experiments and publish in the peer-reviewed scientific literature. Environmental groups don't. But because of disparity of viewpoints and predispositions, the battle lines for biased debate on the environmental impacts of GM plants are clearly drawn. The green groups are anti-GM, while the industry scientists have a perfunctory obligation to be pro-GM. However, if the latter are too pro-GM and release the new GM kudzulike kind of monster into the environment (if ever allowed by regulatory agencies), the backlash would smash markets and profits, render employees jobless, and trigger executive firings. Industry is required to temper its enthusiasm for biotechnology as the science dictates it should. It would be intuitive that environmental groups would allow scientific findings about biotechnology risks and benefits to also shape their agendas. For example, it makes sense that environmentalists would encourage the deployment of a GM plant that makes pest control more environmentally harmonious. But it seems that industry and environmental leaders lock horns in endless battles. Objectivity dissipates. Caught in the middle of the debate are academic scientists employed by universities. Several surveys show that this middle group has enormous public trust and respect; this group is also the one expected to provide the most unbiased interpretations of experimental data. I realize in my simplified survey of the landscape I am leaving out important groups such as regulators, government scientists, and those scientists who work mainly as independent consultants. These all typically fall into the disciplinary sections described below and have approaches and biases most like the respective discipline of their counterparts in universities.

University scientists interested in the environmental impacts of GM plants can be divided into three camps that compose a functional continuum: molecular and cellular biologists, applied agricultural scientists, and ecologists/environmentalists. My descriptions are far from perfect, but they do come from an insider who has worn all of the above hats at one time or another. I've found that most university biologists have no strong opinions on GM plants, since they spend their professional time worrying about other things—this is true even for scientists that work on plants. To a degree, this situation mirrors society at large. That said, most biologists fall into what we generally call "molecular types" or "ecological types," and then there are agricultural scientists. As the biological disciplinary differences have grown stronger, many university departments of biology have split into two different departments during the past 20 years as molecular biology has grown more powerful. Most biologists have at least a rudimentary understanding of the biotechnology involved in producing transgenic plants. They realize that biotechnology is greatly responsible for many of the drugs and crops that are on the market in the United States. They do, however, have diverse interests and do not lose as much sleep over the GM plant controversy as I do.

Before I discuss the scientists, let me mention that there are numerous nonscientist academicians (maybe more of these than scientists) who weigh in on the subject. Typically, they are economists, sociologists, and ethicists who all seem largely unconcerned about the scientific research behind biotechnology, but other aspects of GM agriculture compel them to state their opinions on the matter. They are often more closely allied with environmentalists than biotechnologists; they focus on political and sociological outcomes rather than on biology and science. Rightly or wrongly, I mainly attribute this alliance to the "soft" nature of their work. In fact, many universities have created a new "environmental studies" undergraduate major or minor that typically includes little science. The only problem with black-boxing science in various aspects of the debate is that science is squarely integrated in biosafety considerations; it is part and parcel of the reality of agriculture. The pitfall the nonscientist idealist often falls into is full of media-driven scientific sound bites—often false.

Molecular and cellular biology, agricultural science, and ecology are the three disciplines doing scientific research that provide the greatest impact on agricultural biotechnology. And knowing what these scientists say and who is saying it is important. One would like to think that objective scientific data are decoupled from the scientist or field of study, but it is often helpful to understand the frame of reference or the lens through which each scientist views biotechnology. The following section is a perusal through the disciplines. Sometimes it is tough for a nonscientist to know who the players are because there is no scorecard in science or trading cards for biologists. Baseball has an edge on science in this regard. One indicator to disciplines is departmental affiliation.

The first discipline to be discussed is molecular and cellular biology. Biologists studying at this level best understand the genetic and physiological basis of biotechnology—the wiring and machinery of plants. Some of the opponents of biotechnology insist that this knowledge and involvement in molecular biology makes the molecular biologist unduly biased. I disagree. When bacteria were first genetically engineered in the early 1970s, molecular biologists were the group initially concerned about biosafety. They were apprehensive about the possibilities of genetically engineered superbacteria created in the lab that could have increased invasiveness or reproductive capacity—the parallel of today's situation in plants. Their caution led to a research moratorium until the issues could be resolved. Their concern led to the now-famous conference on the subject held in 1975 at the Asilomar conference facility near Monterey, California. At this conference, the scientists who best understood the science placed it under a microscope for dissection. Their meeting was unique and fruitful in that it paved the way for the National Institutes of Health (NIH) to issue biosafety guidelines for working with recombinant

DNA. Other agencies took the NIH's lead, and regulations from other agencies followed. In the early days of genetic engineering, this extreme caution was entirely appropriate; the science was brand new. That said, today we laugh at the pictures of scientists in moon suits spraying entirely benign antifreeze bacteria onto strawberries in one of the first field tests of GM bacteria. We laugh because we know the noninfective bacteria were entirely crippled and would not survive in nature. Molecular biologists understand the manipulations that create GM plants and have seen firsthand that the minute genetic changes do not make fundamental changes in the nature and ecology of most engineered plants. A GM soybean looks and grows like an ordinary soybean plant, not like a weed. Molecular and cellular biologists are generally supportive of biosafety regulations (although they may occasionally grumble that they are either too restrictive or take too much of their time for paperwork).

Molecular biologists might reside in departments of biology, botany, genetics, or in agricultural departments, such as crop science or plant pathology. Where biology departments have split, they could be in a department called (not surprisingly) molecular and cellular biology. They are generally pro-GM but are much more likely to rely on scientific data more than social or philosophical agendas in making scientific judgments. Their scientific shortcoming is that they often have a limited understanding of ecology and agricultural practices.

The second group of interest is composed of agricultural scientists who, by definition, practice in the more applied and traditional fields of agriculture. They accept modern agricultural systems that include GM plants as a fact of life, and they are the scientists who best understand agricultural practices and the practicalities of agriculture. These are the scientists I most often seek as valuable collaborators because they best understand the relevance of transgenes in particular agricultural systems, and thus they provide ground proofing to what could be otherwise esoteric research. This group is typically pro-GM because of the importance of transgenic traits in production agriculture. They know firsthand about the benefits and continually hear farmers raving about how helpful biotechnology has been to their farms. As a group, agricultural scientists also have the most thorough understanding of economic and ecological benefits GM plants can confer. They typically have some understanding of molecular biology and ecology but don't practice either at the depth of the other two groups. Agricultural scientists typically reside in agriculture departments such as crop science, horticulture, entomology, and plant pathology. Not surprisingly, many come from farms and know how farms work. Although they may be somewhat biased toward pro-GM stands, I've found this group to be quite balanced and practical. Agriculture is populated more with pragmatists than ideologues.

The third group of scientists is ecologists. Most ecologists don't seem to spend much time thinking about molecular genetics or genetic modification. Nonetheless, these are issues that seem to be garnering more interest in the field, and the Ecological Society of America has issued a position statement about GM organisms in 2001. This statement could be described as a middle-of-the-road essay recommending caution and additional research, but at the same time endorsing GM technologies as potential solutions to agricultural and environmental problems.[4] That said, ecologists, who typically reside in biology or ecology departments, have generally eschewed GM plants as potentially interesting research subjects. There are a few excellent ecologists who are involved with GM plant research, but not enough; most ecologists, it seems (albeit from anecdotal evidence and informal conversations), do not consider agricultural systems worthy of their research expertise. There are, of course, notable exceptions to this statement. But as a rule, instead of studying the ecology of a soybean field, ecologists would much rather be off in some "natural" setting doing "real" field research (an Alaskan lake is a more compelling study site than an Iowan corn field), pristine environments are preferable to being down on the farm. These "pristine environment" ecologists aren't very excited about agriculture. And philosophically, these ecologists are also typically real environmentalists: deeply concerned about preserving pristine natural environments, biodiversity, global warming, endangered species, and so on. These issues shrouding the environment often go beyond scientific data and are, at the same time, sometimes antiagriculture.

When we take scientific worldviews into account, agricultural pro-GM agricultural scientists and molecular biologists are often pitted against anti-GM ecologists and environmentalists, such as the Chapela debacle described in chapter 5. Clashes are generally soft, however. These groups stereotypically have had differing views as to the best course of action for the world and the role that biotechnology should play. The agriculturalist as humanist values technology that might help farmers economically and also help farmers better clothe and feed the people of the world. The environmentalist is adamant about environmental preservation and restoration and is not as interested in agriculture. There are inherent philosophical differences, however, when it comes to scientific objectivity about

GM plants. Scientists, molecular, agricultural, and ecological, are closer together in their viewpoints than to those of activists. This fact stems from scientists relying more heavily on scientific data rather than on philosophy or on other nonempirical issues. Science demands objectivity, while environmentalism does not—at least, not institutionalized or dogmatic environmentalism. The scientist is likely to value the environment as well as improved agricultural efficiency. Scientists generally see no dichotomy in preserving wild lands and biodiversity while simultaneously improving the genetics of crops. Why must we choose between biotechnology and the environment, especially when one can help the other?

What Environmental Groups Say about Themselves and about Biotechnology

Once a cause is institutionalized, the next step is to crystallize doctrines and mission. While this is mostly a book about science, in this chapter I want to go beyond the science to analyze extraneous factors that have shaped how people might view biotechnology and perceptions of its safety. One way to do this is to look at sample doctrines and rhetoric. What do environmental groups say about themselves and their values? Are there differences among groups, or are they united in their views? Are they equally scientifically savvy, or is science more of an afterthought?

When I taught an undergraduate entry-level course on biotechnology, I provided students with a list of Websites of environmental groups and agricultural companies. Their challenge was to sort between science and rhetoric, the truth and lies, at least as far as the science would allow them. I asked them about which Websites had the most scientific credibility and which were most persuasive, the goal being to sift through the good, the bad, and the ugly. This general approach seemed to be helpful for my students, and although I will not make any comparisons here with any companies, it should prove valuable to now let activist organizations speak for themselves. I hope we can move beyond institutional and pat rhetoric toward a degree of scientific sophistication. Indeed, objective rhetoric should line up with our scientific knowledge about biosafety. And if it does not, it should be reasonable to ask why.

Every green group that I researched has a Website, and all the Websites had links to genetic engineering. Following is a sample of green groups that ranges the gamut of scientific dialogues.

Union of Concerned Scientists

The Union of Concerned Scientists (UCS; http://www.ucsusa.org/) is, of the groups that I list, the most scientifically erudite. Margaret Mellon and Jane Rissler both communicate frequently to scientists. Mellon has a science Ph.D. and a law degree, Rissler also has a science Ph.D., and both are well respected among environmentalists and scientists alike. Rissler and Mellon wrote a useful short book on the subject of GM plant ecology that was probably the best single work on plant biotech risk assessment when it was published in 1996.[5] Since that time, many of the issues and fears raised in their book have been assuaged with data.

Of all the groups surveyed, UCS seems to be the most adept at focusing on the current scientific literature. But its real forte and niche seem to be as a regulatory watchdog, akin to the Consumer's Union. When I reviewed the UCS Website, the most interesting information was on the procedures EPA had used in the re-registration of Bt crops in 2000. During that year, the monarch butterfly/Bt corn pollen brouhaha was still in full swing, and the suite of *PNAS* papers were not yet published. In 2000, the jury was still out as well on the effect of Bt on monarch butterfly. UCS closely monitored the EPA's movements, wrote letters to the agency, and continually questioned the regulatory decisions made about continuing to allow Bt corn to be grown in the face of apparent risks to the monarch. After the *PNAS* papers were published in 2001, it became quite clear that Bt corn was not a significant risk to monarch butterfly populations. The last time I checked (in 2002), UCS had not yet updated its Website to reflect the new data and current opinion. Nonetheless, it seems to play an interesting role in keeping balance in the biotech debate.

ETC Group, Formerly RAFI

RAFI, Rural Advancement Foundation International (http://etcgroup. org/main.asp), is the group most famous in biotechnology circles for brilliantly renaming the technology protection system "Terminator" then following up with a closely related "Traitor" moniker. It executed a very effective public relations campaign to nip that particular technological development in the bud. People accepted RAFI's argument that such a technology could really hurt poor farmers by forcing them to purchase engineered seeds from large, multinational agrochemical companies. As discussed earlier, Terminator likely would not really change any paradigm in seed purchase practices in the developed world and would have likely never been used in subsistence farming anyway. Now RAFI has a new name but the same cause. The letters ETC in the new name stand for erosion, technology, and concentration. On its Website, the ETC group defines its mission, which is geared toward the protection of food and agriculture for the developing world. Under "technology" it says this about biotechnology: "ETC group focuses on the social and economic impacts of new biotechnologies. ETC group is not fundamentally opposed to genetic engineering, but we have profound concerns about the way it is being foisted upon the world." It goes on to conclude that biotechnology is neither safe to people nor the environment. While I can't foresee divorcing biotechnology from free markets anytime soon, it seems to me that ETC group would be happy if that were the case. It seems that this divorce would be a prerequisite for ETC group's endorsement. And perhaps ETC group has a grand plan for implementation that does not require corporate development and marketing, but it seems to me that without corporations and patents, there would be no GM plants, since most public entities, such as university research groups, don't have the funds or drive to develop transgenic crops for mass plantings of millions of acres. It is the unique role of private enterprise to fuel technological development and commercialization, a role not endowed to universities and government. ETC group, while claiming to not be philosophically opposed to biotechnology, seems to be practically against the implementation of GM plants used in today's agricultural world.

Greenpeace

Greenpeace (http://www.greenpeace.org/homepage/) is now a multinational environmental organization that focuses on everything everywhere. Its fight against genetic engineering is only one of Greenpeace's many campaigns. As it points out on its Website, "Greenpeace opposes all releases of genetically engineered organisms into the environment. Such organisms are being released without adequate scientific understanding of their impact on the environment and human health." Greenpeace claims that genetic engineering is inherently risky, insisting that there is a lack of knowledge on long-term effects of GM plants. On its Webpage about Bt transgenic plants, it systematically, albeit briefly, discusses the high-profile studies that are covered in the middle part of this book. Although the discussion references scientific publications and has a purposeful tone of being rigorous, many papers that do not bolster the Greenpeace view of the world are not cited. For example, it ignored the weight of evidence from the suite of the monarch *PNAS* papers that proved that monarch populations would not be harmed by Bt corn pollen. It also doesn't acknowledge any environmental benefits of GM crops.

Sierra Club

The Sierra Club (http://sierraclub.org/) is the oldest of the environmental groups reviewed here. Its Web report on genetic engineering updated in 2001, was decidedly opposed to genetic modification of any type. Its Webpage states, "Genetic engineering now poses a very grave threat to the natural environment." This statement and corollaries seem overly one sided in scope and justification. The Sierra Club claims we are at a crossroads: either we stop genetic modification now, or we'll have a genetically modified planet. Its members obviously believe that once transgenes are released into the environment, a point of no return will be passed; furthermore, the case is made that GM invaders will obliterate nature as we know it.

The Sierra Club's discussion of Bt corn adversely affecting soil microorganisms included no citations or data to support its conclusions. It claims that since biotechnology advocates (industry and governmental regulators) have no willpower or ability to control GM plants, then it is up to the Sierra Club to police the science and commercialization processes. At one point, the Sierra Club's report takes on a quasi-religious tone, as it waxes poetic about John Muir (its founder) and his spiritual quest seeking and communing with nature. I was befuddled as to the relevancy of the diatribe to genetic engineering and found myself wondering where the committee that wrote the report got its information. Then, suddenly, it all made sense with this statement: "There is not a shred of evidence that John Muir would have regarded the release of genetically engineered organisms to the environment with anything but shock and outrage." They essentially hold the mantra, What would John Muir do? The Website report earnestly urges the Sierra Club troops to fight against the evil infidel that is the "powerful corporate forces" incarnate. It seems to me that the information included in the Website leads one to believe that genetic engineering cannot be redeemed, no matter what environmental benefits might be in the offing. In the continuum of activist organizations, the line has been crossed from where science has a place in environmentalism to a predomination of quasi-religious dogma.

Earth First! and Earth Liberation Front!

I group these last two organizations together because they represent the more militant faction of the environmental movement. The Earth First! Webpage (http://earthfirst.org/) has a monkey wrench icon predominantly displayed; this seems to represent its approval for the practice of "monkey wrenching," destroying private property and machinery that could be used by people they believe will harm the environment. Targets might include foresters or developers who could alter present environmental states. While the Webpage claims that Earth First! doesn't organizationally endorse monkey wrenching, it says that some of its members might believe differently and act accordingly. It espouses deep ecology as a spiritual force and indeed, there is significant religious dogma on its pages.

Finally, Earth Liberation Front (ELF) (http://www.earthliberationfront.com/) has admitted to numerous acts of vandalism and ecoterrorism. For example, it has taken credit for bombing the construction site of a biotech building at the University of Minnesota in 2002. It has a Web-based report on all its criminal activities, and it states that it is impossible to contact the ELF ward in any geographic area, as it is secret society. Even though one of the ELF's guidelines precludes injury to "animal, human, and non-human," these biota seem to be at collateral risk while ELF is doing what it does naturally.

Thus, not all green groups are alike. They range from mild to wild. But they virtually all are against genetic engineering and generally all for organic agriculture as the world's answer to food production. The surface has only been scratched here in this brief survey, but policies and beliefs of the myriad activist groups are redundant with the organizations just discussed. Clearly, scientific data exonerate biotechnology from blanket guilt as dangerous for the environment. But in the activist equation, GM plants compose only one side. The other side is the glorious benefits and sustainability of organic agriculture. How good are organics, and can organic farmers feed the world? What if they used biotechnology?

Organic Agriculture

Where does a nostalgic fairyland view of agriculture end and science begin? The old adage is that the degree that farming is romanticized is inverse to how close one lives from a farm. I recall sitting in a conference hall in

London hearing Benny Haerlin, a leader in Greenpeace, espousing the benefits of organic agriculture, and natural foods. While heralding organic agriculture as the answer, he failed to mention any of the inefficiencies inherent to organic production or its unsustainability as the exclusive form of farming. The fact remains that to even attempt to feed the world using organic agriculture, we would need to convert current wild lands to agricultural fields, thus disrupting biosphere function and wiping out entire ecosystems.

I used to be an organic farmer. In fact, before 1940, almost all American agriculture could be described as organic. It was also low yielding, inefficient, and precarious. And while I grew up on my grandparents' farm in the 1960s in North Carolina, one could say that my farming family was somewhat behind the times on farming technology. We were all organic farmers. Yes, we had tractors and various farm implements, barbed wire and electric fences, but I think that was about the end of technological achievement on the farm. I vividly remember sitting high upon a pile of manure on a trailer behind a tractor, where I pitchforked putrid organic matter onto fields. We called this procedure fertilizer application. I recall picking Japanese beetles off various horticultural crops: crop protection. I killed rats with a frog gigger in the egg production facility, better known as the cageless chicken house. This was recreation. Sometimes the rats prevailed over my efforts, and the chickens' quality of life (and egg production) suffered. My memories of organic farming are mixed. There are a few good recollections, but mostly thoughts of "so this is why there is value in education" dominated my childhood. As an adult I am free now to join the ranks of organic farmers, but I'd like to think I've already had a lifetime's worth of experience of sitting on piles of cow manure. It seems to me that most advocates of organic farming as *the* agricultural solution to world hunger are relatively rich American suburbanites or city dwellers who have not had the joy of first-hand experience on an organic (or any other) farm. There are millions of people in developing countries who are not organic farmers by choice, but as a byproduct of poverty and the lack of technological assistance.

My cloistered experiences aside, there is a growing market of people who wish to pay more for organically grown produce, and, as a result, organic farming is booming in the United States. Still, only about 0.3% of America's vegetables come from organic farms, and I have serious doubts that the buyers of "100% organic" really know what they're paying more for.

There are limits on the productivity and viability of extensive organic agriculture. It is simply too land hungry. Scientists have written some interesting articles on the need to embrace technology to save wilderness.[6,7] It was estimated that in 1993 about 36% of the earth's surface was used for agriculture. If agricultural technology development had ceased in 1961, the estimated figure would be closer to 63% of the earth's surface that would have to be farmed to provide for the world's need to feed and clothe people. The reason we need less land is mainly attributable to genetic improvement of crops (the green revolution), better protection from pests (yes, chemicals), chemical fertilizer (not manure), and irrigation. Without the implementation of these technologies, more rainforests would have had to be cut down and land plowed under. Crop yields are not increasing from year to year now as fast as they did during the green revolution. Yet additional yields can be expected to be realized in the future by tweaking genetics, especially in the crops grown in the developing world.

Transgenic insect- and disease-resistant varieties would greatly improve crop protection, which would result in less reliance on chemicals and better yields. Unfortunately, organic farming does not tout any effective methodologies for increasing crop yields. It is an accepted fact that without chemical fertilizer applications, the limiting factor to plant growth is most often nitrogen deficiency. Yields are severely limited without supplemental fertilization to provide essential mineral nutrients. Because organic farmers use only manure as fertilizer, yields are suboptimal compared to fully fertile conditions found on conventional farms. There is simply not enough crap to fertilize all crops in this world. Organic agriculture seems to be practiced by subsistence farmers without choice and by farmers in relatively rich countries to provide a type of food as a luxury item to those who can afford it. It is unconceivable that organic production could ever feed the world.

In one of the most comprehensive studies comparing conventional and organic agriculture, the yield problem of organics was confirmed.[8] The study was performed over 21 years in what has been called one of the most fertile valleys in Switzerland. The researchers reported that crop yields were 20% lower in the organic plots compared with conventional. For potatoes, the yields were only 58–66% of those grown in conventional systems. Part of the decrease in yield for potatoes was the result of low potassium and uncontrolled disease by the pathogen that caused the great Irish potato blight, *Phytophtora infestans*. Although the authors of the study concluded that organic farming is a realistic alternative to conventional farming, it is obvious that they would have needed at least a third more land to produce the same number of potatoes as those produced using conventional methods. But it is interesting to note that if the technology pool had been increased to include genetically modified potato varieties, the yield would have been markedly similar to conventional methods, while the insecticide inputs would have been significantly less.

Given a choice, most people, I think, place a higher value on environmental stewardship than on a crop production system based on ideology. Most reasonable people are not anxious to see more wild lands converted to farms. Nor do they wish more people to starve to death. Most people agree that all safe technological tools ought to be brought to bear to make farms as efficient and productive as possible. Even though it seems I have personally rejected organic methods, my tiny home garden is largely organic—but mainly due to neglect. For several years I even used fish feces as fertilizer for my tomatoes. But then again, my gardens have always been for fun. Observations of agricultural biodiversity are intriguing to someone not gardening for optimal yield or needing to produce pristine, marketable fruit and vegetables. I am amazed to observe huge tomato hornworms snacking on spare tomato plant foliage. But I doubt the subsistence or even professional farmer would share my pastime or even have the luxury of watching potential food crops being gobbled by obese worms. If insects eat my quasi-organically grown tomatoes, I'll have to go shopping. If third-world farmers suffer crop failures, then they may starve. If we can deliver insect- and disease-resistant crops to farmers in the developing world, along with all biosafety features and precautions, they stand a chance of having a significantly better quality of life.

GM Organics Anyone?

The intrinsic inefficiencies of organic agriculture are obvious. Couple this unsustainability of organics with the glaring fact that the production of a Bt protein to kill insects on a GM plant fits well with organic practices of not using chemical insecticides. Organic farmers have been using Bt bacterial sprays for years. I recognize that the notion of ever growing a Bt transgenic plant is unfathomable to today's organic farmer; nonetheless, there is no good biological justification for this out-of-hand rejection. Genetic modification could be an organic tool to increase efficiency of farming with few to no chemical inputs. When the USDA recently established organic certification guidelines, the organic farming community overwhelmingly rejected the inclusion of GM plants in its methodology. Why? Most of the reasons were ideological or those related to marketing strategies. In fact, I have never heard of any scientific rationales for organics omitting GM technologies. What is the difference between a GM plant that confers insect resistance by expressing a Bt gene and spraying Bt onto crops? Both methods avoid the need to spray synthetic chemical insecticides, but the Bt plants are better for the environment because no fuel is used and no pollution is produced by the spraying process needed to deploy organic insecticides.

There could be some valid economic reasons having to do with marketing and product identity as a result of how the public views organic products. But marketing and public perceptions are intertwined with complexities that often go beyond science. Some people believe that organically grown produce is more nutritious, but there are no data to substantiate that view. The choice to be an organic grower or eater has more to do with ideology than the environment or nutrition.

Parting Thoughts

True environmentalism values wild lands and sustainable agriculture. I've yet to talk with someone who does not believe affordable and abundant food for all people is a worthy goal. Biotechnology can help fulfill this goal. We need to move beyond combative rhetoric and look toward science as a focus for meaningful dialogue. We are quickly learning about the biosafety of GM plants, and the news so far is good. Biotechnology should not be viewed as an agricultural cure-all, but it is certainly not the demon many organizations make it out to be either. Although farming is not optimally environmentally friendly, it seems that biotechnology is helping move it into a greener position; GM plants having far more environmental benefits than risks. It should be intuitive that agriculture can be optimized by genetic improvement of crops. But leaving agricultural problems behind us for a while, are there any biotechnological developments on the horizon designed to explicitly seal environmental fissures? The answers compose some of the most exciting science and technological developments of the twenty-first century.

12

Futurama

Greenetic *Engineering for a Greener Tomorrow*

Thus far, this book has largely concentrated on the most often discussed, proven, or potentially negative environmental impacts of GM plants. But the environmental benefits of today's GM crops are astounding and overwhelming. Thirty-eight trillion GM plants grown in the United States have improved the environment directly and indirectly. One direct benefit, for example, is the decrease in pesticide applications. Less crop dusting has been needed to control insects, and so each year millions of gallons of insecticide have not had the opportunity to leach into lakes, streams, and groundwater. The indirect benefit of herbicide-tolerant crops is reduced soil erosion. In many places, before no-till agriculture was practiced, the plow would kick up opaque curtains of dust as topsoil was translocated off the farm and into water and other undesirable places. No-till agriculture was facilitated largely by glyphosate, sold as Roundup herbicide (now that the patent has elapsed, several companies sell glyphosate weed killers). And herbicide-tolerant crops have encouraged no-till agriculture.

While there are clearly environmental benefits to today's GM plants, they were not invented or marketed to help the environment per se but largely to assist farmers in operating more efficient and economical farms. Today's row crops are merely the first wave of many GM plants that will be environmentally beneficial. There are several GM plants in the research and development pipeline that are designed to help solve environmental problems and will be forthcoming in significant numbers, but they are not yet on the market. Many of these have little to nothing in common with agricultural crops.

Myriad environmental applications will be discussed in this chapter, but no attempt will be made to exhaust the list of research in progress. I only want to give a flavor of things to come. My favorite environmental objective involves deploying transgenic plants designed to detect buried land mines and then subsequently clean up the explosives that remain as toxic soil pollutants. This use of transgenic plants in land mine cleanup, as well as other phytoremediation applications for sequestering and/or detoxifying heavy metals and other pollutants will no doubt help the environment by returning polluted or otherwise human-altered ecosystems to more natural states. These and other examples will be discussed in this chapter. Phytoremediation and phytosensor plants are merely the tip of the environmental improvement iceberg. Scientists are currently performing research to engineer trees with less lignin, grains for chicken feed that will yield less nutrient-rich guano (a water pollutant when in surface runoff), and GM plants that make biodegradable plastics and substitutes for fossil-fuel—biorenewable materials and fuels.

It is a mystery why any dedicated environmentalist would oppose these research efforts and the subsequent development and commercialization of biotechnological tools designed especially for environmental improvement. Certain activists dedicated to a certain ideal will no doubt continue to find objectionable points to biotechnology as a means to an end, but pragmatic environmentalists won't care about the means as long as the environment benefits in the end. In this chapter I describe research efforts that are not merely environmentally friendly as by-products but that are specifically aimed for the sole purpose of improving environmental quality.

Plants to Detect Land Mines and Degrade Explosives

There is an emerging research area for GM plants for the purpose of what I call phytosensors.[1] Theoretically, phytosensor plants can be used to detect various specific impurities in the environment and signal the presence, and perhaps amounts, of pollutants on site and in real time. Plants could be engineered to continually monitor contaminant movement. Plants are renowned for their capabilities to respond to environmental cues by turning gene expression on and off, thereby altering innate physiological and biochemical reactions. For example, they are known to possess heat-, cold-, light-, pathogen-, and salt-inducible gene promoters, which in turn regulate the expression of specific genes that help cope with specific stresses. Plants also have signal transduction pathways, sometimes similar to those found in animals, which allow them to perceive and respond to hazardous chemicals in their environment. (After all, they cannot run away from environmental hazards and toxins). The genetics and biochemistry of plants can be manipulated to produce phytosensors that can serve as unmanned and natural sentinels for reporting on the movement and types of pollutants in the environment. GM plants might be the ideal sentinels to monitor wide geographic areas in real time. We are only beginning to understand the breadth of environmental conditions and pollutants that plants can respond to using their native genetics and physiology.

One of the simplest systems that could be engineered into plants would be the fusion of a specific inducible promoter to a reporter gene and then placement of this construct into a plant of interest, such as a tree, that could be planted to continually monitor and report on environmental quality. One reporter gene that has been used extensively is one coding for green fluorescent protein (GFP) originally isolated from a jellyfish (figure 12.1). GFP simply fluoresces a bright glowing green under a black light. When engineered into plants, GFP also makes plants fluoresce, and it is a key technology for producing sentinel plants. In the case of phytosensors, we want the fluorescence be switched on only when the contaminant is present. My research group is investigating GM plant fluorescence using GFP for all sorts of applications, and we've worked with engineers to develop detection instrumentation as well.[2] My lab is producing GM plants that will give specific fluorescent signals in response to a host of substances of interest: plant diseases, chemical warfare derivatives, and explosives, to name a few. Plants that are responsive or sensitive to explosives could someday be used to detect land mines. If we are successful, the outcome, as one soldier recently told me, will revolutionize the way warfare is done. My dream is not to alter warfare, though, but rather for people to use plants in humanitarian demining operations. Subsequently, perhaps the same GM plants will perform dual duty to phytoremediate the residual explosives from the soil, thus, returning former minefields to cornfields.

Land Mine Phytosensors

It is an understatement to say that land mines are a huge environmental and humanitarian travesty. Since World War II the deployment of land

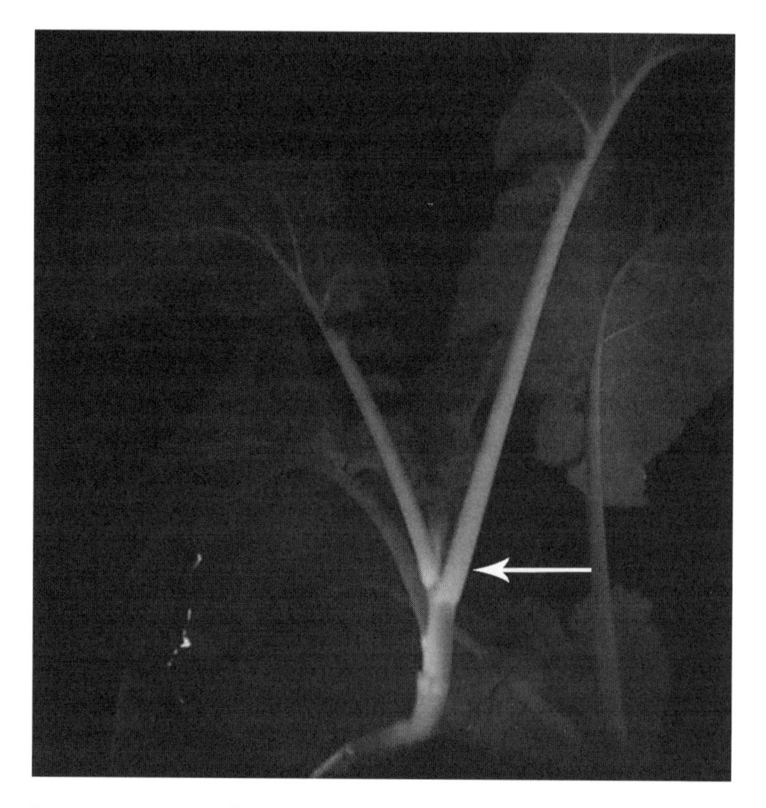

Figure 12.1. Canola transformed with a green fluorescent protein gene under a UV light (arrow). In the background to the right is a non-GM plant that does not fluoresce green. (Photo by Matt Halfhill.)

mines has been a deplorable part of warfare in that civilians, not military personnel, are the people most often injured by mine explosions. The United Nations considers the land mine problem to be a major global challenge (www.unfoundation.org); it is estimated that somewhere between 60 and 100 million land mines are deployed in at least 70 different countries. The number of noncombatants killed or maimed is estimated at 26,000 per year. Approximately one-third of these casualties are children.

Most land mines contain high explosives such as trinitrotoluene (TNT) and/or royal demolition explosive (RDX). Modern mines are cheaply produced (~$10 or less) and have plastic housings that leak explosives. Antipersonnel mines are about the size of a hockey puck. Metal detectors cannot be used to find land mines with plastic housings, the typical encasing material. State-of-the-art detection includes dogs trained to sniff explosives; however, since trained dogs are not readily available in large numbers, people trained to slowly prod the ground to feel out the location of mines is the typical mode of operation in humanitarian demining. Needless to say, there is room for methodological and technical improvements to locate mines. Lots of money has been poured into chemical sensing electronic noses (e-noses) and other technology for finding buried land mines and explosives, but little success has stemmed from these efforts. It has proven to be a technologically challenging task.

In 1999, I first heard about some Army-sponsored work in the use of genetically modified bacteria for the detection of explosives. In this project, researchers sprayed bacteria on an experimental minefield to map out the location of individual mines. The microbial cultures were composed of GM bacteria engineered with a toluene-inducible promoter fused to the GFP gene. The research team hoped that the bacteria would fluoresce green when they came in contact with the soil located over the land mines. This scheme was somewhat effective, but the bacteria also fluoresced when they came in contact with plants near and not so near the buried land mines. The plants had taken up the TNT and concentrated breakdown compounds in their tissues, and the bacteria sometimes responded to explosive now in plant leaves. The system was less than perfect, and it seemed that it had certain constraints that would prevent converting this technical development to one that could be used effectively in live minefields. It came to mind that we might not need the bacteria if we could make a plant that could detect TNT directly and fluoresce using essentially the same types of molecular mechanisms. It seemed quite desirable to delete the need to spray GM bacteria over wide geographical areas. GM bacteria would be considerably more controversial than growing GM plants. Could you imagine a foreign country inviting Americans, perhaps associated with the military, to spray genetically engineered bacteria (albeit harmless strains) on their plants and soil located on dozens to thousands of military-sensitive acres? In general, I think the contentious perceptions of growing GM plants pale in comparison to the use of GM bacteria. Many bacteria cause disease. and a few have even been used in biological warfare. In contrast, plants are purposefully grown in our gardens and we eat them for breakfast, lunch, and dinner.

It became clear that plants might be the ideal biological candidates to detect explosives; it was known that plants absorb TNT through the roots and naturally transform it to other compounds. Plants also respond to TNT by altered growth and physiology. Although TNT and other explosives are toxic to plants, they are much more poisonous to animals. My lab has gone into a genomic fishing expedition in the model plant arabidopsis (figure 12.2) to find plant genes that are turned on by TNT and then clone them to drive plant fluorescence, so it can be detected from a distance using laser-based instrumentation.[3] The end goal of the research is to produce a plant that lights up when it is growing on ground covering a land mine (figure 12.3). The results of our research are promising: TNT-inducible promoters have been discovered and are being characterized. Funding agencies are supporting the research at the university, and there is good potential of forming a spin-off company for the purpose of developing biotechnology tools for humanitarian demining, as well as other phytosensors.

But finding and removing the buried land mines is just the first step. After mines are removed, explosives remain in the soil. Although explosive-ladened soil is not in danger of blowing up, high explosives such as TNT are toxic to wildlife and humans. One of the great travesties of land mine deployment is that valuable agronomic land is taken out of production. TNT in foodstuffs is poisonous and will hasten sickness and death, albeit more slowly than being blown up. Likewise, in addition to mines, there is unexploded ordnance (bullets, shells, bombs, etc.) that needs to be detected and cleaned up. GM plants should be useful here as well.

Phytoremediation of Explosives

Unlike phytosensing, a relatively new concept, phytoremediation, has been around for a few years. Several start-up companies, such as Edenspace Systems Corporation, have been formed for the purpose of using plants to remove pollutants from soil and water. After plants sequester toxins in green tissue, they can either be subsequently extracted from harvested plants or degraded into compounds that are less toxic or nontoxic *in planta*. Why phytoremediation? As stated before, plants are very good at responding to their environment and coping with environments with harsh components. Some plants are somewhat suited for phytoremediation in a non-transgenic state, but genetic engineering has the potential to drastically improve the effectiveness and efficiency of the phytoremediation processes and systems.

Neil Bruce in England and his colleagues have produced plants that express a bacterial nitroreductase gene that speeds up a plant's ability to degrade TNT to toluene breakdown products.[4] Both TNT and the vari-

Figure 12.2. *Arabidopsis thaliana*. The genome of this plant has been sequenced. Its short life cycle and small genome have made it the fruit fly of the plant world. (Photo by Reginald Millwood.)

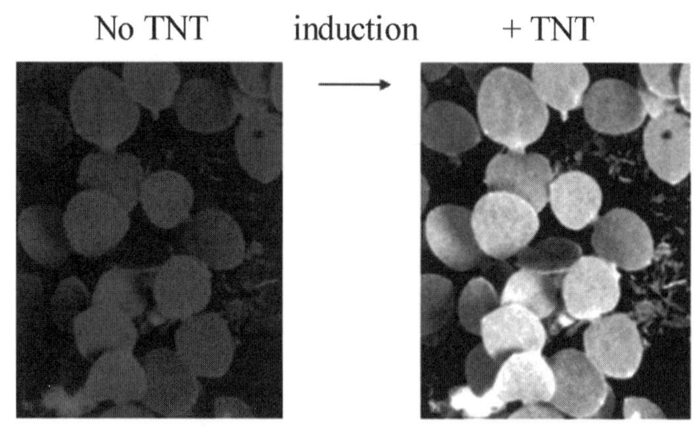

Figure 12.3. Phytosensor scheme. If plants are engineered with a construct consisting of a TNT-inducible promoter fused to green fluorescent protein (GFP), they will fluoresce when exposed to explosives under a UV light.

ous toluene-containing products are toxic. They can cause liver damage and cancer. Nonengineered tobacco plants naturally transform toluenes to inert by-products. Genetic engineering was used to increase the speed of detoxification of TNT, thus perhaps making the pytoremediation of explosives more practical. So we might imagine a phytosensor plant and a phytoremediation plant as a single entity delivered to Afghanistan or Cambodia in one large case of seeds that can be used to both find land mines and then subsequently remove explosives from the soil, thereby transforming minefields into fields that are more peaceful and nourishing.

Phytoremediation of Other Environmental Pollutants

Heavy metal contamination of soils is a problem throughout much of the world. Like land mines, the problem is human caused. Technology should be leveraged to aid in the cleanup of toxic metals such as copper, zinc, nickel, cadmium, lead, chromium, and mercury. Each year, billions of dollars are spent to decontaminate badly polluted sites using the lowest of low-tech means: excavating metal-laden soils and trucking them to some desolate spot for burial. Current technology simply transports the problem to another locale, albeit in a less densely populated area. Other low-tech methods, such as incineration, also fall short of the mark in addressing the real problems in an environmentally friendly manner.

Many plants and microbes have peptides and proteins that can bind and/or detoxify metals. These include metallothioneins and phytochelatins. The presence of these and other compounds confers tolerance to metals that allow survival of the host organisms. Research is being performed to assess the effects of these phytoremediation candidates in GM plants. One transgenic plant-based phytoremediation strategy is to simply upregulate genetic elements that are already present to some degree in plants to make them more effective in uptake, tolerance, and bioaccumulation of metals. Another tool is genetic modification to introduce novel metal-binding elements. It seems that nature has given us tools that are sometimes not in the optimal packages for remediation, and rearranging and swapping these packages is the power of genetic modification. For example, the metal hyperaccumulating herb, *Thlaspi spp.*, is a very tiny plant that can sequester a considerable percentage of its biomass in zinc. Biotechnologists would like, if possible, to transfer *Thlaspi* genes responsible for metal hyperaccumulation into easily cultivated plants that produce large biomass for harvesting (figure 12.4). Some of these larger plants, like Indian mustard (*Brassica juncea*), are already used for

Figure 12.4. *Brassica juncea* is a plant often used in phytoremediation.

phytoremediation in the nonengineered state. Edenspace (www.edenspace.com) calls Indian mustard its phytoremediation champion. However, it is easy to imagine increasing the phytoremediation strength of the champion using biotechnology. Increasing metal uptake and tolerance twofold could remove metal in half the time—in fact, Indian mustard has already been engineered for metal tolerance by Colorado State University researchers.[5]

Another strategy for producing plants that can accumulate and then detoxify metals has been pursued at Rich Meagher's lab, where GM plants accumulated, transformed, and then volatilized mercury.[6,7] This project is perhaps the best known GM phytoremediation application to date because of its advanced state, and the high quality of the research, which used a sophisticated metabolic engineering approach. Mercury exists in soils as a pollutant in the toxic methylmercury form or as ionic mercury, which is less toxic. Meagher's group engineered plants to express a bacterial *merB* gene that codes for a protein that converts mercury from methyl to ionic form. Furthermore, the researchers reengineered plants with a bacterial *merA* gene that encodes a protein that converts ionic mercury to elemental mercury, which is the least toxic form of all. Under high temperatures, elemental mercury is volatilized and dissipates from the plant. Therefore, a transgenic poplar tree engineered with two *mer* genes could be grown at a mercury-laden site for decades while removing the toxic metal from the soil.[8] The number of metal phytoremediation GM plant projects underway is staggering. For a current review paper of the full implications of phytoremediation of metals, see Krämer and Chardonnens.[9]

There are numerous other pollutants, such as trichlorophenols and PCBs, that could be degraded and cleaned up by GM plants. The promise of using plants for on-site removal of organic toxins lies in the feasibility of converting toxic compounds to harmless substances such as carbon dioxide, ordinary salts, oxygen, and water. Genetic modification is truly the bioremediation grail in that the genetic diversity of metabolic degradation pathways in microbes and fungi might be readily transferred into plants that could be planted on site and monitored for many years, until the mess is suitably cleaned up without destroying the soil ecosystem.

Ligninless Trees

Trees are one of our most valuable renewable resources and the primary biomass source of numerous ecosystems. Well into the foreseeable future, forest trees will continue to provide building materials for housing and furniture, as well as pulp and paper. The environmental problem with the pulping process is that lignin, an elementary component of wood, must be extracted from pulp to ensure that paper is of a certain quality. Perhaps even the most fanatical of environmentalists would not tolerate the use of lignin-loaded toilet tissue. And there is copious lignin in wood: up to one-third of the biomass of trees is lignin, so its extraction is not trivial. Chemical extraction is environmentally costly because it often uses large quantities of toxic substances, including bleach, to manipulate pulp for the production of paper. The biotechnological answer to the lignin problem is the manipulation of the lignin biosynthetic pathway in the tree. It is a complex biochemical problem because the pathway is variable among types of trees. Nonetheless, some successful early results indicate that producing trees with reduced lignin is a reasonable goal if GM techniques can be used. Both poplar and pine species have been genetically modified for reduced lignin content.[10] Although it is too early to know exactly how well a ligninless tree will grow and how strong the trunk and wood might be, the GM approach is promising enough that millions of research dollars have been invested already. If GM trees do live up to their potential in reducing lignin, there will be two significant environmental benefits. First, as alluded to above, there will be a tremendous reduction of toxic chemicals now needed in pulp and paper processing, a significantly decreased environmental problem. Second, pulp yields will be increased on a per-log basis, in turn decreasing the number of acres of trees that will need to be grown per unit paper production. Engineering trees for less lignin will greatly improve the environmental health of areas surrounding pulp and paper mills.

Phytase Chicken Feed

Phosphorus is a required element for all organisms, but it can be a pollutant in the wrong place and in high amounts. Plant seeds (grain), which are the most important raw ingredients in animal feeds, store much of their phosphorus as phytate. Unfortunately, phytate consumed in grain is not readily digestible and absorbed by farm animals, and therefore it is destined to reside in piles of manure. It's bad enough that animals cannot benefit from the potential phosphorus nutrition, but the nutrient can end up in water as a pollutant.[11] This indigestible phytate thereby poses a twofold environmental problem. First, more inorganic phosphorus must be mined from the ground to supplement phosphorus nutrition in animal feed. Second, and more important, the runoff of phytate into streams and rivers causes algal blooms, which can, in turn, cause significant fish kills. Nutrient leaching from chicken and pig farms into aquatic ecosystems has long been pegged as a problematic point source for phosphorus pollution. The dual nutritional/pollution problem has been addressed to some degree by supplementing animal feed with enzymatic phytase isolated from a fungus; in the animal's digestive system, phytase breaks down phytate, thereby increasing phosphorus bioavailability and increasing the animal's nutrition. Quite simply, more efficient absorption of phosphorus in animals would greatly reduce pollution and eutrophication of waterways. The phytase supplement approach has proven effective, but it also increases the cost of animal feed. To decrease costs of delivering phytase into animal guts, a transgenic approach could be useful. A Dutch research group cleverly expressed the fungal phytase gene in transgenic plants to high levels.[12] Subsequently, other researchers have been investigating the feasibility of expressing the phytase gene in feed seeds such as soybeans. When phytase-transgenic soybeans were fed to chickens, their guano contained 50% less inorganic phosphorus than when fed non-transgenic feed.[13] If soybean, corn, and wheat seeds can be engineered to successfully deliver phytase on a routine basis in feed, farmers will have solved both animal nutrition and collateral environmental damage problems simultaneously.

Plastic Plants and Biofuels

The last type of environmental application we examine here involves using plant-based biomolecules as a substitute for petroleum-based products. Because plants are an important sustainable and renewable resource for energy production, they are attractive as miniature factories to produce bioplastics and biofuels on the farm. Several research groups have investigated the production of organic polymers such as polyhydroxyalkanoates (PHAs) and polyhydroxybutyrate (PHBs) in transgenic plants. Since the early 1990s, researchers have worked on methodologies to transfer bacterial genes into plants for the production of biodegradable plastics.[14] The research's logical end is producing plastics in the field using GM plants rather than using microbial fermentation devices to manufacture biodegradable plastics. These plastics could be an alternative to traditional petroleum-based products. Petroleum-based plastics are neither from a renewable resource nor biodegradable. The research has been marching forward to the degree that even Monsanto has invested in research into plastic plants.[15]

In the same mold as plastic plants, transgenic petrol plants could be used for biomass or biofuel production. These products could perhaps at least partially replace gasoline and diesel fuel for internal combustion engines as well as coal in electrical generators. For many years, corn has been converted to ethanol for mixing with gasoline. Plant-produced oils from oilseed plants have also been used as a biodiesel fuel source. There is currently much discussion about the role of genetic modification to increase biomass production or to fine-tune lipid metabolism for environmentally friendly fuels. At present, the available plant products are not favorable on an economic scale and also suffer from suboptimal composition constrained by native plant metabolism. Altered genetics can help remove both metabolic constraints and perhaps increase economic viability by the creation of better biofuel plants. We are beyond the halfway point in fossil fuel depletion, and alternatives are needed. Even if we were not running out of petroleum, everyone agrees that the dependence on fossil fuels is indelibly linked with environmental accidents during the drilling and shipping of crude oil. A long-term solution will no doubt involve renewable energy sources, including biofuels, which will have transgenic plants at their core.

As indicated above, an important impediment to growing plants (GM or not) for biodegradable plastic and biofuel production is economics. The extraction costs currently associated with the low amounts of plastics produced by a GM plant are greater than traditionally produced plastics. Likewise, the price of a gallon of cooking oil or ethanol for drinking or therapeutic purposes far exceeds the price that is competitive for a gallon of biofuel. At some point, as biotechnology advances and the price of petroleum increases, an economic crossover point will occur that will favor economical production of plastics and biofuels in plants. Of course, there are also political reasons to move more quickly toward the adoption of biofuels. I can envisage governmental subsidies for biofuels and biofuel research and development to ease reliance on foreign oil. Biotechnology is a critical factor in biorenewable energy development.

When "Genetic Pollution" Rights Ecological Wrongs

Other ecological applications of biotechnology will crop up that beg for a paradigm shift in our perceptions of the role of GM plants in nature. Currently, the paradigm demands exclusion of transgenes from nature; transgenes might be tolerated on the farm, but not outside it. I predict that two test cases in which GM plants will ultimately be considered welcome guests in nature will involve curing dogwood anthracnose in the common dogwood and chestnut blight in the American chestnut. Although both of these new GM disease-resistant plants would alter the present "natural" ecological interactions existing today, I predict they will be eventually accepted as desirable interventions into the natural world for the sake of ecorestoration. Of course, such acceptance will arrive following dialectics to overcome preconceived notions dealing with "genetic pollution." Today opponents of biotechnology refer to genetic pollution and gene flow synonymously, and, of course, it is taken to be necessarily environmentally harmful. But these two biotechnological alterations would, in fact, serve to correct downturns in the populations of these two highly important, naturally occurring, species in the eastern United States. Biotechnology could remedy these two drastic, human-caused, ecological maladies in which one tree species has been wiped out and the other is in serious decline. In both scenarios, wild trees would be genetically engineered, and gene flow from the engineered plants to other wild trees of the same species would be the desired end point. (Would we call them "genetic supplements"?) As transgenes move into tree populations inhabiting natural ecosystems, host plant fitness will increase because of enhanced disease resistance. These GM plants will not interbreed with any species other than their own, and therefore the transgenes will be contained, albeit in the natural distribution of these two species in the United States.

The only parties that could conceivably oppose these releases of transgenes into nature would be those with philosophical objections that would equate this application as irrevocably destroying nature. As with most of the controversies commingled with misperceptions about biotech risks covered in previous chapters, philosophy will overshadow scientific facts and benefits in many minds. How we view nature takes us back to the beginning of this book. Those who see nature as a wild place with genes flying to and fro quite naturally will accept one more gene or so as pleasant supplements in these ecologically important trees. Those who view nature as a tidy, organic box where any technology is anathema will fight the introduction of GM ecorestoration trees. So, in many ways, it will be business as usual for GM plant scientists who have become accustomed to hullabaloo. While ecorestoration would be the desired end point for all parties, genetic supplements could be easily viewed by some (especially initially) as genetic pollution. Perhaps now is the optimal time for the reader to judge whether GM dogwood and chestnut would be appropriate as vehicles for ecorestoration or not—while the head is full of scientific knowledge about biotech ecology (the point of this book), and also while the heart can be cool in considering what is not presently technically feasible. Dogwood and chestnut ecorestoration currently exist only in the hypothetical realm.

Dogwood

Dogwood, *Cornus florida*, is a beautiful, naturally occurring small tree that inhabits the east coast of the United States. Economically, it is a very popular ornamental tree used in landscaping, especially in the Southeast. A terrible fungal disease, dogwood anthracnose, was first noticed in the late 1970s on dogwoods in the northeastern United States, and a similar disease popped up that infected the dogwood species (*Cornus nuttallii*) in the Pacific Northwest. In the late 1980s, it was clear that the disease had spread to the southern Appalachian Mountains, where it now affects every county in the region while simultaneously becoming increasingly more devastating to natural dogwood populations. Dogwood anthracnose kills leaves of lower branches of trees, and then the lower branches themselves (figure 12.5). Trees become disfigured with time and sometimes die. The problem is especially bad at higher elevations (over 1000 m) and in shady, damp places. While some disease-resistance genes have been found in dogwood populations and in related species, the beautiful ornamental is at risk of going the way of the American chestnut in certain habitats.

Is it possible and desirable to engineer dogwood in the lab to have full resistance against dogwood anthracnose? Would disease resistance increase the competitiveness of the species? If a disease-resistance gene such as one encoding chitinase could be engineered into dogwood and the plants could be released into the environment, not only could the trees grown as horticultural ornamentals be saved, but wild-growing dogwood could also be spared additional decline as the result of anthracnose disease.

Figure 12.5. Symptoms of dogwood anthracnose disease. (Photo by Mark Windham.)

Dogwood populations would eventually rebound to pre-anthracnose levels. Because the disease is most likely a very recent and exotic introduction, it is difficult to imagine any negative ecological side effects from allowing a resistance gene to sweep through native populations. The transgene spread throughout wild dogwood would save this American treasure from demise and return dogwood to its original ecological stability.

Chestnut

Another even more dramatic example of how genetic engineering could bring ecological restoration to the Appalachian Mountains is to revive the American chestnut, *Castanea dentata*, using modern biotechnology. In the early 1900s, the American chestnut, a formerly dominant tree in the Appalachian Mountains and surrounds, was discovered to be diseased with cankers, soon recognized as chestnut blight. The causal agent of chestnut blight, a fungus, was found growing unchecked under the bark and around the stem, and it finally killed the entire above-ground portion of the tree. The blight is believed to have been introduced into the United States by the importation of ornamental Japanese chestnuts (*Castanea crenata*) that resist the disease but still carry it. During the 20-year period following the introduction, the dominant American chestnut was reduced to an understory sapling, which, even today, resprouts from ancient root stocks as multistemmed shrubs (see figure 4.10). Before the saplings are large enough to reproduce, they are decimated, once again, by blight that is still resident in the tree. Substantial research has focused on the disease-causing fungus. Fungal strains with decreased virulence have been isolated and studied. The underlying molecular cause of virulence is becoming better understood. At the same time, the arduous task of breeding chestnuts for resistance (crossing resistant Japanese trees with susceptible American trees) and selection for resistant trees has begun.

A genetically purer route of developing resistant trees would be to genetically engineer resistance directly into American chestnut so that the other "foreign" (Japanese chestnut) genes could be excluded. The disease-resistant transgenic American chestnut version would have only one or two foreign genes rather than the thousands introduced from Japanese chestnut. The GM American chestnut could then be released into its native environment, where it would spread back into the Appalachians to resume its rightful role as a dominant species, the ecological niche it held until the early twentieth century. Of course, in the process of chestnut population restoration, it would inevitably displace dozens of hickory and oak species, tulip poplars, and ash trees in the Appalachians en route to resuming its historically natural ecological role. Today's forests would be drastically changed to a status unknown to any of us living today. Thus, biotechnology may be used in ecorestoration to correct a mistake made 100 years ago that disrupted the natural balance of the Appalachian biome. Turnabout would indeed be fair play. I'm sure that arguments will emerge to the effect that biotechnology might ruin natural ecosystems. But I see it as one technology fixing another's mistake and taking ecosystems back to an equilibrium enjoyed in the nineteenth century.

How realistic are these scenarios? Neither dogwood nor chestnut is an easy species to engineer, and they have not been successfully engineered to date. However, tissue culture systems have been established, and I have no doubt that both species will be made transgenic given enough time and effort. Likewise, there is no magic disease-resistance gene that has been identified as a silver bullet for these problems. But I predict that both technical hurdles will be addressed for these trees and their diseases, as well as for Dutch elm disease and those debilitating diseases that infect a suite of other important tree species.

As I finish writing this chapter, I am listening to the crickets and frogs making delightful noises from my back deck overlooking woodlands and pond. I could close on this note of genetically engineered ecorestoration and be completely satisfied, but it would not really be fair because I would have left out a few key realities on the genetically modified planet. For instance, the world of GM plants is greatly impacted by corporate agendas. We've seen already that these plans are moderated by regulatory agencies that are the watchdogs of public good. Finally, there is a little philosophic muddy water that still needs some straining. I hope to give the reader a few missing pieces and wrap up with a holistic picture of the future of GM plants in light of their past.

13

Conclusion

Out of Right Field and into Home

Joltin' Joe and Relays

As the deep sacrifice fly ball lands in the right fielder's glove, the runner on third base tags up and heads for home. Few major league outfielders have had a strong enough arm to reach the catcher in a single bound for the throw-out. Indeed, Joe DiMaggio was a rarity. The dilemma that the typical outfielder is struck with is to decide whether to attempt a throw-out at home or try to hit the relay man.

Plant biotechnology has been mired in similar dilemma. Many parties have tried to throw the long ball out of right field for home and have gone errant. Indeed, plant biotechnology has been the prototypical confusing subject out of right field. It has been fraught with misunderstandings and confusion from day one. What has been needed for some time in biotechnology is what is known in baseball as a relay man—the player who receives the throw into the infield, who in turn makes an accurate short throw to home. If the throw comes in time, the runner coming from third base might be tagged out. I hope that this book serves as a relay man of sorts—one that renders an easy-to-catch and accessible message. One key element toward any success in this realm is the accurate placement of agricultural biotechnology in an appropriate context of today's farming techniques. This concluding chapter is meant to wrap up the discussion of the environmental impacts of GM plants and to reiterate contexts of modern farming. The chapter can be summed up with three maxims in no particular order: things are not always as they seem; good guys and bad guys don't always wear the appropriately colored hats; and it is impossible to separate agriculture from nature.

The Convolution of Agriculture and the Environment

Less Is More

Conventional wisdom holds that it is undesirable to lose farmers to other jobs. Likewise, more companies are better than just a few that function as an oligopoly. However, as fewer and fewer farmers and agricultural companies fulfill agricultural services, several favorable developments have resulted. It can be argued that fewer, yet more interconnected players might be good for the environment. Modern farmers are linked with agricultural companies that, in turn, control most of the technology. There are also fewer secrets in agriculture anymore. Syngenta knows what Monsanto, Dow AgroSciences, Bayer, and DuPont (the big five) are up to and vice versa. Technological reciprocity and parity seem to be on the upswing as well. So what does this have to do with the environment? The small world of agriculture is ever-shrinking with demands for greater environmental stewardship and cooperation.

All the companies are under increasing social and regulatory pressure to provide environmentally benign products. Large companies are financially able to comply with increasing regulatory requirements—for registration of products to compliance. Increasing regulation has selected for fewer and larger companies, and because of this, regulations can also be further increased with minor corporate hardships.

Farmers are under similar pressure to tread lightly on the environment. Both regulation and technology have selected fewer farmers who can adapt to changing environments. In turn, these few flexible farmers can then adopt newer technologies—and not just biotechnologies. For example, one aspect of conventional farming has born a new approach called "precision farming."

Precision Farming and Changing Landscapes

Precision farming uses satellite and global positioning technology, chemistry, ecology, and biotechnology to direct the use of herbicides and other pesticides, along with fertilizer, only where needed in the field. It is quite sophisticated and scientific compared with the traditional and romanticized farming. In the age of chemical dependence, a crop duster might fly over vast acres of land and dispense, in one fell swoop, dozens to hundreds of gallons of toxins that would saturate a field. In contrast, precision farming uses remote sensing via satellite imagery downloaded to a global positioning unit on a tractor to inform the farmer exactly where a fungicide application is needed. So, the modern farmer can save money, fuel, and chemicals by treating only the area needed. One of our phytosensor applications is plants that detect their own diseases and then respond with a genetically engineered fluorescence signal, which is envisioned to be useful someday in precision farming.[1]

Romance, Chemicals, and Modern Science

Romantic farming suffered from low yields and hardships, but it was environmentally friendly. Chemical-dependence farming was characterized by dramatically increased yields, but it carried an environmental cost. Modern, scientific farming has the best of both worlds. But there are forces that seem to be slightly stuck in the chemical era (even scientists), when it was chic to bash agriculture.

In that chemical-era context, agricultural sciences and ecology have had opposing philosophies and motivations that seem to be disintegrating. In my experience, most ecologists became interested in their field because of their concern for, or even love of, the environment. Obviously, agricultural scientists are primarily focused on improving farming and farm output. The old assumptions are that agriculturalists don't care about the environment and that ecologists are inherently hostile to conventional farming. The assumptions made by both of these kinds of scientists about one another are often being dismissed as they see that in a shrinking world, agriculture and the environment are increasingly intertwined and codependent. The environmentalist increasingly understands that to save existing wild lands from agricultural conversion, farming must be made more efficient to sustain net yield. The agriculturalist is learning how to minimize environmental impacts to encourage sustainable agriculture. Environmental stewardship has become a key component in agriculture. Paradigms are changing, but not fast enough. There are still some misgivings and misunderstandings about fundamental perceptions about the contexts of agricultural practices and nature: if they exist among scientific disciplines, how much more mystification must cloud the general public. A relay man is needed.

Conscious and Subconscious

Most people agree that it is appropriate for vast amounts of arable land to be dedicated to producing food. However, they are not keen about converting additional area from wild lands to agriculture. Most people place high value on existing wild lands and want them maintained as such; many people even desire existing farms to remain intact and not converted to suburbs. But people universally do not want farms to negatively impact their personal environment or wild lands. In short, they would like to receive all the benefits farming has to offer (e.g., food and fiber), but otherwise enclose farms into an indelible bubble. But is this realistic? We consciously know that the world is composed one continuous landscape (albeit landforms often separated by large bodies of water), but subconsciously we tend to compartmentalize our personal environments (house, yard, neighborhood, and city) as one world, farms existing on another planet, and vacation spots including wild lands on yet another world. And I think in most of our minds' eyes, we subconsciously picture farms as quaint Norman Rockwell paintings, complete with silos, wooden barns with faded paint, and 1920s-era farmhouses. In one thought we welcome this vision into our landscape. But information about toxic chemicals and now mysterious biotechnology taint the Norman Rockwell picture, and we build walls to compartmentalize agriculture even more—farmer as bubble boy. But science clearly shows that biotechnology is environmentally advantageous. Feelings and science don't always mix well.

Nonetheless, attempting to freeze technology and squeeze reality into some bygone romantic era will always be frustrating to everyone concerned. I find activism that ignores real-world needs and developments very frustrating at several levels. In the modern era, many assumptions surrounding agriculture and the environment are being proved false. For instance, environmentalists believe farming is always inherently environmentally destructive.

Salton Sea

A curious example of farming being environmentally advantageous is the case of the Salton Sea in southern California. In this case, agricultural runoff has greatly increased bird biodiversity around the sea.[2] Located just to the east of San Diego County, the Salton Sea was accidentally created when a levee broke in the early twentieth century, allowing the Colorado River to fill a 360-square-mile depression. Today it is filled solely with agricultural runoff from the rich farms of the Imperial Valley. This situation has resulted in the creation of an extremely salty body of water in which few aquatic organisms are fit to dwell, but the birds love it. Approximately 400 species of birds live around the Salton Sea, making the body of water the second-highest site of bird biodiversity in the United States. Included here are the endangered brown pelican and Yuma clapper rail. The Salton Sea has been called "the crown jewel of avian biodiversity."[2] However, a threat to this newfound biodiversity is, ironically, diminished farming and agricultural runoff into the lake. Water rights are forcing farmers to irrigate less. As a result, there is less nutrient-rich runoff filling the lake, and so the Salton Sea is becoming smaller and saltier and decreasing aquatic food choices for birds. Some hard decisions will have to be made. But, clearly, this is one example of where farming and agricultural runoff have served an interesting and valuable ecological function. This example may seem anomalous, but it reinforces the notion that natural, environmentally, and aesthetically valuable are not synonymous. The Salton Sea bird biodiversity is valued by environmentalists, even if the cause is traditionally not embraceable by the activist community. With environmental health, there are always constraints and real services that must be considered. Agriculture is an ecosystem service that must be provided to feed a growing global population and provide a decent quality of life. We also need a sustainable environment. It is a false sense of choice to either champion agricultural or environmental causes exclusively—we must have both. Agriculture companies, among others, have awakened to these facts and place these conditions as vital to their existence. They have no choice really, because the world has no choice for sustained living. Biotechnology plays a key role.

Corporate Agendas

In chapter 11, I made quite a big deal about environmental activists' agendas. It would seem that equal time should be paid to those of agricultural corporations—the big five. However, that is simply not necessary, or even possible. Agricultural corporations have a boringly simple agenda: profits. Companies make a profit by delivering viable products to their clients (farmers) that will allow them to sustain profits as well. These products are conventional seed, chemicals, seed treatments, and GM plants. In the future, they might also sell information for precision agriculture. Farmers and markets determine the products and services. Technology will always play a role in the types of products, but the agricultural scenario at any time will be exercised in a world in which the amount and quality of food as well as the environment are considered together. As we've seen, all these pressures are squeezing farmers. While farmers are growing more grain on less land with fewer inputs, there are environmental constraints posed by regulatory officials and public pressures. In the context of environmental stewardship, some compression is not bad, but the bursting point for farmers in the United States and other developed countries is close at hand. People living in suburbs that are encroaching on farmland are complaining more of offending odors and sounds. Overregulation increases costs. Some farmers worry that trade wars might make GM crops difficult to sell. And now there is the new threat of agricultural terrorism that is hanging like a dark cloud. It is a tough but a technologically exciting time for farming. It is clear that biotechnology has a helpful role in farming because it increases productivity and environmental stewardship, but GM crops are helpful only if there is consensus for acceptance of these new tools.

Risks and Benefits of GM Plants in Context

This book has mainly focused on agricultural GM crops and their environmental impacts; most are positive and a few are negative. The GM crops of tomorrow will continue this trend, but, on occasion, there may be significant environmental risks that need to be assessed. We must have the science and public funding to work through these snafus before new GM crop varieties go to market. It has worked so far. In addition to new crops, plants that are engineered for environmental clean up and specific environmental applications will be developed and deployed. As I advised the New Zealand Royal Commission, saying no to genetic modification via plant transformation will be akin to saying no to jet engines for intercontinental air flight.

After growing trillions of transgenic plants in the United States during the past 14 years, there have been no ecological disasters, no injuries or deaths, no GM crops invading natural ecosystems, and no negative measurable effects whatsoever; yet, surprisingly, GM plants have not been universally accepted in agricultural systems on the global level. And there is no doom on the horizon that I can see. GM plants have proven to be among the safest of agricultural technologies. Tractors flip, farm workers have been poisoned by pesticides, and now environmental extremists vandalize fields believed to contain GM plants. Farming is an inherently risky endeavor. David Pimentel and Peter Raven state that there are 110,000 nonfatal pesticide poisonings reported each year in the United States and 10,000 cases of cancer attributable to pesticides.[3] In contrast, not only has no one been harmed by GM plants, they are systematically replacing risky pesticides.

GM plants do not add to any of these existing risks or pose any new compelling risks of any magnitude, yet they are highly regulated—more regulated than any other farm technology. While GM plants have one or a few well-characterized gene inserts and largely predictable and measurable transgene expression levels that lead to expected phenotypes, they are subject to extraordinary federal regulatory scrutiny. At the same time, plants derived from wide crosses as well as mutagenesis through chemical or radioactive bombardment are not regulated at all. These non-GM plants have thousands of new and randomly changed genes and are truly genetic black boxes. I know of no efforts to regulate the growing of such mutagenized crops in the United States. And we know much less about the genetics and biochemistry of mutagenized crops compared with GM varieties. GM plants are the most deeply studied and understood (genetically, physiologically, and ecologically) plants ever grown anywhere.

Risks don't start and end with field crops. Water tables are increasingly being decreased in areas of heavy irrigation. Irrigation is not regulated. Suburban development and sprawl is eating natural areas and farmland across America. These activities are not regulated by federal officials, and it seems people only get concerned when wetlands are involved. However, there is only a finite level of land to build outward. Eventually we have to infill.

Golf courses, likewise, convert multiuse land to monocultures of turfgrass that are dedicated to people with the resources and discretionary time needed to hit a tiny white ball for hours on end. Yet the chemical and energy inputs necessary to sustain pristine, clipped, creeping bentgrass greens and fairways are enormous and have negative environmental impacts as well. Golf courses are not regulated by the EPA or the USDA.

While known pests are excluded from import to the United States by USDA, invasive exotic horticultural plants such as Japanese privet,

Japanese honeysuckle, and wisteria are planted and grown across the country by homeowners and landscaping companies at their discretion. We know that these landscaping plants all escape cultivation and disrupt natural ecosystems, and obscure goals (figure 13.1). Horticulture was the route of initial escape of kudzu. Importation of exotic horticultural varieties that, absent from natural enemies, can run amok outside of managed landscapes is a known hazard. These invasive exotic plants that spread into unmanaged ecosystems are completely unregulated, yet we know they can be hazardous to the environment. Every reader of this book has ob-

Figure 13.1. An exotic invasive, albeit horticultural, plant, Japanese honeysuckle, has covered the author's basketball goal during the writing of this book.

served instances (whether or not the species were identified) of horticultural plants turned weedy and gone wild.

So should all these environmental hazards and technical black boxes continue unabated? Should these land-use activities and horticultural plants be more regulated to put them on par with GM plants? The answers are no and no, respectively, with a few caveats.

Regulations

There is a need to holistically revisit issues dealing with land use, exotic and invasive plants, and sustainable agriculture. GM plants are not the hazard they've been made out to be. We truly need to eliminate invasive horticultural landscape plants and be more careful to avoid exotic introductions. We need to be better stewards of the environment. But overregulation in agriculture is placing, and will continue to place, additional financial strain on an already low-margin business that is absolutely necessary to sustain human societies. Farmers are near the breaking point as a profession. Does that mean we need to regulate GM plants less? I think so. Currently, the USDA regulates nonrisky GM plants to the same degree as potentially risky transgenic plants. There are no exceptions in the USDA, no carte blanche for any GM plant. There is definitely overkill on some of the regulations, and, I might add, their implementation. There are some new biotech products, such as GM plants that produce pharmaceutical proteins, that require intense regulatory scrutiny. But there is no need to regulate all GM plants because research plainly shows that there is nothing inherently environmentally risky about the biotechnology used to produce them. The risks will always lie with the types of transgenes and the choice of crops. For example, crops engineered with domestication genes should be regulated to a much lower degree than plants producing pharmaceutically-active proteins.[4] GM plants are definitely overregulated with regard to their known and suspected risks; a one-size-fits-all strategy is not appropriate. The regulations are strong in the United States because of perceived unknown risks with a new technology. The technology is no longer new, though, and it will be downright ancient in 20 years. Maintaining the current amount of regulatory control of GM plants as a whole by then will be a foolish waste of resources with regard to what we can currently predict from the science. Perhaps it is even foolish now to spend so much money controlling such safe entities. Thirty-eight trillion transgenic plants grown with no incidence of problems add credence to this statement. Certainly America is wasting money on regulatory redundancy.

I believe that the United States should look closely at the Canadian system, which focuses on novel phenotypes and traits produced in plants, and not the means by which they were produced. In the Canadian system, regulatory scrutiny is tripped for herbicide-tolerant plants because of the novel trait, whether the plants were produced through mutagenesis, wide crosses, or GM technology. Under such a strategy, novel plants, such as those being used for pharmaceutical production, would receive high amounts of regulatory scrutiny, while well-characterized traits that are no longer truly novel would be accepted as safe. There is certainly room for maturity and growth in agriculture; technology will be important to the safe and stewardly production of food and fiber. Genetic modification that is safe for humans and the environment will be an important tool and will almost certainly play an important role in ecorestoration. I hope that scientific data and appropriate interpretations prevail.

Concluding Thoughts

Let's revisit baselines, timelines, and points of comparison. I believe we should leave romantic comparisons in the past. We cannot bring back the old days of subsistence farming using 100% organic methods. The only relevant choices revolve around today and tomorrow. Will any new technology at any particular juncture make agriculture more productive and safe? Will it make the environment cleaner? Baseline scientific data are important to track changes, and we should let science override nostalgia in our future decisions.

The real danger in resting on science and technology, in my opinion, is when we view any new technology as the silver bullet and adopt it as an end-all to any specific problem. Americans have a fabulous history of doing one thing, then temporarily closing the book on problems until new and sometimes worse problems resurface. I know people whose one thing for nutrition is fast-food burgers. Americans' transportation one thing is the automobile. We've seen the era come and go where the insect control one thing was chemical insecticides. While there is nothing inherently hazardous about genetic modification, there is a hazard of making a few types of GM plants our new one thing. It is simply intuitive that planting crops tolerant to a single herbicide year after year and then spraying that herbicide year after year is going to lead to resistance problems. In this area, the global agricultural community can learn wisdom from our organic farming comrades. There is value in diversification in cropping systems. Planting monocultures of genetically identical crops and using the same pest control system will lead to long-term sustainability issues. Once an insect is resistant to Bt Cry1Ac toxins or a weed is resistant to glyphosate, unless there is a genetic or physiological cost for harboring that resistance gene, the resistant insect or weed will always be with us. If this book does nothing else, I hope it defocuses us from peering down the narrow pipe that represents the risks of growing GM plants and toward the larger issues of farming and land use, which will undoubtedly have GM plants as vital components in the long term.

While the tools we use to build the edifice are important, they are not of ultimate importance—the edifice is. Visitors to Agra may be quite curious of the craftsmanship and tools that went into building the Shah Jahan's mausoleum, but when faced with the splendor of what could be the most wonderful building in the world, one is in awe of the Taj Mahal itself, not chisels and hammers. Biotechnology has its role to play in agriculture, but it is no silver bullet. Nor will it be an end-all in ecorestoration. It is just a tool. If we use GM plants wisely, they will be important tools to bring us closer to building sustainable agricultural systems, which will help to both feed the world and improve the environment. The elegance and structure of sustainable agriculture is more beautiful than any human edifice in that it sustains human life.

References

Chapter 1

1. Ho, M.-W. 1998. *Genetic Engineering: Dream or Nightmare?* Gateway Books, Dublin, p. 1.

2. Crawley, M.J., R.S. Hails, M. Rees, D. Kohn, and J. Buxton. 1993. Ecology of transgenic oilseed rape in natural habitats. *Nature* 363:620–623.

Chapter 2

1. Holm, L., J. Doll, E. Holm, J. Pancho, and J. Herberger. 1997. *World Weeds: Natural Histories and Distribution*. John Wiley & Sons, New York, p. 674.

2. Baker, H. 1965. Characteristics and modes of origin in weeds. In *Genetics and Colonizing Species* (H.G. Baker and G.L. Stebbins, eds.), pp. 147–168. Academic Press, New York.

Chapter 3

1. Klein, T.M., E.D. Wolf, R. Wu, and J.C. Sanford. 1987. High-velocity microprojectiles for delivering nucleic acids into living cells. *Nature* 327:70–73.

Chapter 4

1. Warwick, S.I., M.J. Simard, A. Légère, L. Braun, H.J. Beckie, P. Mason, B. Zhu, and C.N. Stewart. 2003. Hybridization between *Brassica*

napus L. and its wild relatives: *B. rapa* L., *Raphanus raphanistrum* L. and *Sinapis arvensis* L., and *Erucastrum gallicum* (Willd.) O.E. Schulz. *Theoretical and Applied Genetics* 107:528–539.

2. Adam, D. 2003. Transgenic crop trial's gene flow turns weeds into wimps. *Nature* 421:462.

3. Hall, L., K. Topinka, J. Huffman, L. Davis, and A. Good. 2001. Pollen flow between herbicide-resistant *Brassica napus* is the cause of multiple-resistant *B. napus* volunteers. *Weed Science* 46:688–694.

4. Warwick, S.I., M.J. Simard, A. Légère, L. Braun, H.J. Beckie, P. Mason, B. Zhu, and C.N. Stewart. 2003. Hybridization between *Brassica napus* L. and its wild relatives: *B. rapa* L., *Raphanus raphanistrum* L. and *Sinapis arvensis* L., and *Erucastrum gallicum* (Willd.) O.E. Schulz. *Theoretical and Applied Genetics* 107:528–539.

5. Stewart, C.N., Jr., J.N. All, P.L Raymer, and S. Ramachandran. 1997. Increased fitness of transgenic insecticidal rapeseed under insect selection pressure. *Molecular Ecology* 6:773–779.

6. Stewart, C.N., Jr., M.J. Adang, J.N. All, P.L. Raymer, S. Ramachandran, and W.A. Parrott. 1996. Insect control and dosage effects in transgenic canola, *Brassica napus* L. (Brassicaceae), containing a synthetic *Bacillus thuringiensis cryIAc* gene. *Plant Physiology* 112:115–120.

7. Ramachandran, S., G.D. Buntin, J.N. All, P.L. Raymer, and C.N. Stewart, Jr. 1998. Movement and survival of diamondback moth, *Plutella xylostella* (Lepidoptera: Plutellidae) larvae in mixtures of non-transgenic and transgenic canola containing a *Bacillus thuringiensis cryIA*(*c*) gene. *Environmental Entomology* 27:649–656.

8. Ramachandran, S., G.D. Buntin, J.N. All, P.L. Raymer, and C.N. Stewart, Jr. 1998. Greenhouse and field evaluations of transgenic canola against diamondback moth, *Plutella xylostella* and corn earworm, *Helicoverpa zea*. *Entomologia Experimentalis et Applicata* 88:17–24.

9. Ramachandran, S., G.D. Buntin, J.N. All, B.E. Tabashnik, P.L. Raymer, M.J. Adang, D.A. Pulliam, and C.N. Stewart, Jr. 1998. Survival, fitness, and oviposition of resistant diamondback moth (Lepidoptera: Plutellidae) on transgenic canola producing a *Bacillus thuringiensis* toxin. *Journal of Economic Entomology* 91:1239–1244.

10. Stewart, C.N., Jr. 1999. Insecticidal transgenes into nature: Gene flow, ecological effects, relevancy and monitoring. In *Symposium Proceedings No. 72, Gene Flow and Agriculture—Relevance for Transgenic Crops* (P.J.W. Lutman, ed.), pp. 179–190. University of Keele, 12–14 April 1999. British Crop Protection Council, Surrey, U.K.

11. Ramachandran, S., G.D. Buntin, J.N. All, P.L. Raymer, and C.N. Stewart, Jr. 2000. Intraspecific competition of an insect-resistant transgenic canola in seed mixtures. *Agronomy Journal* 92:368–374.

12. Halfhill, M.D., H.A. Richards, S.A. Mabon, and C.N. Stewart, Jr. 2001. Expression of GFP and Bt transgenes in *Brassica napus* and hybridization and introgression with *Brassica rapa*. *Theoretical and Applied Genetics* 103:659–667.

13. Chevre, A.M., F. Eber, A. Baranger, and M. Renard. 1997. Gene flow from transgenic crops. *Nature* 389:924.

14. U., N. 1935. Genomic analysis in *Brassica* with special reference to the experimental formation of *B. napus* and peculiar mode of fertilization. *Japanese Journal of Botany* 7:389–452.

15. Halfhill, M.D., R.J. Millwood, P.L. Raymer, and C.N. Stewart, Jr. 2002. Bt-transgenic oilseed rape hybridization with its weedy relative, *Brassica rapa*. *Environmental Biosafety Research* 1:19–28.

16. Scott, S.E., and M.J. Wilkinson. 1999. Low probability of chloroplast movement from oilseed rape (*Brassica napus*) into wild *Brassica rapa*. *Nature Biotechnology* 17:390–392.

17. Stewart, C.N., Jr., M.D. Halfhill, and S.I. Warwick. 2003. Transgene introgression from genetically modified crops to their wild relatives. *Nature Reviews Genetics* 4:806–817.

18. Stewart, C.N., Jr., and E.T. Nilsen. 1995. Phenotypic plasticity and genetic variation of *Vaccinium macrocarpon* (American cranberry) I. Reaction norms of clones from central and marginal populations in a common garden. *International Journal of Plant Sciences* 156:687–697.

19. Gressel, J. 1999. Tandem constructs: preventing the rise of superweeds. *Trends in Biotechnology* 17:361–366.

20. VanGessel, M.J. 2001. Glyphosate-resistant horseweed from Delaware. *Weed Science* 49:703–705.

21. Mueller, T.C., J.E. Massey, R.M. Hayes, C.L. Main, and C.N. Stewart, Jr. 2003. Shikimate accumulates in both glyphosate-sensitive and glyphosate-resistant horseweed (*Conyza canadensis* L. Cronq.) *Agricultural and Food Chemistry* 51:680–684.

22. Basu, C., M.D. Halfhill, and C.N. Stewart, Jr. 2004. Wood genomics: new tools to understand weed biology. *Trends in Plant Science* 9.

Chapter 5

1. Quist, D., and I.H. Chapela. 2001. Transgenic DNA introgressed into traditional maize landraces in Oaxaca, Mexico. *Nature* 414:541–543.

2. Stewart, C.N., Jr. 2002. The future of transgenic plants. *ISB News Report* (January), pp. 6–7.

3. Mann, C.C. 2002. Has GM corn "invaded" Mexico? *Science* 295: 1617–1618.

4. Conko, G., and C.S. Prakash. 2002. Report of transgenes in Mexican corn called into question. *ISB News Report* (March), pp. 3–5.

5. Metz, M., and J. Fütterer. 2002. Suspect evidence of transgenic contamination. *Nature* 416:600–601.

6. Kaplinsky, N., D. Braun, D. Lisch, A. Hay, S. Hake, and M. Freeling. 2002. Maize transgene results in Mexico are artefacts. *Nature* 416:601.

7. Quist, D., and I.H. Chapela. 2002. Quist and Chapela reply. *Nature* 416:602.

8. Editorial note. 2002. *Nature* 416:601.

9. Brown, P. 2002. Mexico's vital gene reservoir polluted by modified maize. *Guardian* (April 19). Available: http://www.guardian.co.uk/Archive/Article/0,4273,4397091,00.html.

10. Stewart, C.N., Jr., M.D. Halfhill, and S.I. Warwick. 2003. Transgene introgression from genetically modified crops to their wild relatives. *Nature Reviews Genetics* 4:806–817.

11. Suarez, A.V. 2002 Conflicts around a study of Mexican crops. *Nature* 417:897.

12. Worthy, K., R.C. Strohman, and P.R. Billings. Conflicts around a study of Mexican crops. *Nature* 417:897.

13. Metz, M., and J. Fütterer. 2002. Metz and Fütterer reply. *Nature* 417:897–898.

14. Kaplinsky, N. 2002. Kaplinsky replies. *Nature* 417:898.

15. Pesticide Action Network of North America Website. Available: http://panna.igc.org/about/board.html. Accessed June 6, 2002.

16. U.S. Patent 5,723,765. 1998. Control of plant gene expression. Inventors: M.J. Oliver, J.E. Quisenbury, N.L.G. Trolinder, and D.L. Keim. U.S. Patent Office, Washington, D.C.

17. Kuvshinov, V., K. Koivu, A. Kanerva, and E. Pehu. 2001. Molecular control of transgene escape from genetically modified plants. *Plant Science* 160:517–522.

Chapter 6

1. Carson, R. 1962. *Silent Spring*. Houghton Mifflin, Boston.

2. Ehrlich, P.R. 1968. *The Population Bomb*. Ballantine Books, New York.

3. Charles, D. 2001. *Lords of the Harvest: Biotech, Big Money, and the Future of Food*. Perseus Publishers, Cambridge, Mass.

4. Ewen, S.W.B., and A. Pusztai. 1999. Effects of diets containing

genetically modified potatoes expressing *Galanthus nivalis* lectin on rat small intestines. *Lancet* 354:1353–1355.

5. Stewart, C.N., Jr. 2003. Press before paper—when media and science collide. *Nature Biotechnology* 21:353–354.

6. Losey, J.E., L.S. Rayor, and M.E. Carter. 1999. Transgenic pollen harms monarch larvae. *Nature* 399:214.

7. Cornell University. 1999. Press release. Available: www.cornell.edu/Chronicle/99/5.20.99/toxic_pollen.html.

8. Jesse, L.C.H., and J.J. Obrycki. 2000. Field deposition of Bt transgenic corn pollen: lethal effects on the monarch butterfly. *Oeceologia* 125:241–248.

9. Hellmich, R.L., B.D. Siegfried, M.K. Sears, D.E. Stanley-Horn, M.J. Daniels, H.R. Mattila, T. Spencer, K.G. Bidne, and L.C. Lewis. 2001. Monarch larvae sensitivity to *Bacillus thuringiensis* purified proteins and pollen. *Proceedings of the National Academy of Sciences USA* 98:11925–11930.

10. Pleasants, J.M., R.L. Hellmich, G.P. Dively, M.K. Sears, D.E. Stanley-Horn, H.R. Mattila, J.E. Foster, P. Clark, and G.D. Jones. 2001. Corn pollen deposition on milkweeds in and near cornfields. *Proceedings of the National Academy of Sciences USA* 98:11919–11924.

11. Oberhauser, K.S., M.D. Prysby, H.R. Mattila, D.E. Stanley-Horn, M.K. Sears, G.P. Dively, E. Olson, J.M. Pleasants, W.-K.F. Lam, and R.L. Hellmich. 2001. Temporal and spatial overlap between monarch larvae and corn pollen. *Proceedings of the National Academy of Sciences USA* 98:11913–11918.

12. Stanley-Horn, D.E., G.P. Dively, R.L. Hellmich, H.R. Mattila, M.K. Sears, R. Rose, L.C.H. Jesse, J.E. Losey, J.L. Obrycki, and L. Lewis. 2001. Assessing the impact of Cry1Ab—expressing corn pollen on monarch larvae in field studies. *Proceedings of the National Academy of Sciences USA* 98:11931–11936.

13. Sears, M.K., R.L. Hellmich, D.E. Stanley-Horn, K.S. Oberhauser, J.M. Pleasants, H.R. Pleasants, H.R. Mattila, B.D. Siegfried, and G.P. Dively. 2001. Impact of *Bt* corn pollen on monarch butterfly populations: a risk assessment. *Proceedings of the National Academy of Sciences USA* 98:11937–11942.

14. Zangerl, A.R., D. McKenna, C.L. Wraight, M. Carroll, P. Ficarello, R. Warner, and M.R. Berenbaum. 2001. Effects of exposure to event 176 *Bacillus thuringiensis* corn pollen on monarch and black swallowtail caterpillars under field conditions. *Proceedings of the National Academy of Sciences USA* 98:11908–11912.

15. Hartzler, R.G., and D.D. Buhler. 2000. Occurrence of common milkweed (*Asclepias syriaca*) in cropland and adjacent areas. *Crop Protection* 19:363–366.

16. Losey, J.E. J.J Obrycki, and R.A. Hufbauer. 2002. Impacts of genetically engineered crops on non-target herbivores: Bt-corn and monarch

butterflies as a case story. In *Genetically Engineered Organisms: Assessing Environmental and Human Health Effects* (D.K. Lenourneau and B.E. Burrows, eds.), pp. 143–166. CRC Press, Boca Raton, Fla.

17. Stewart, C.N., Jr., and S.K. Wheaton. 2001. GM crop data—agronomy and ecology in tandem. *Nature Biotechnology* 19:3.

Chapter 7

1. Schuler, T.H., R.P.J. Potting, I. Denholm, and G.M. Poppy. 1999. Parasitoid behaviour and Bt plants. *Nature* 400:825–826.

2. Hilbeck, A., M. Baumgarter, P.M. Fired, and F. Bigler. 1998. Effects of transgenic *Bacillus thuringiensis* corn-fed prey on mortality and development time of immature *Chrysoperla carnea* (Neuroptera:Chrysopidae). *Environmental Entomology* 27:480–487.

3. Hilbeck, A., W.J. Moar, M. Puztai-Carey, A. Filippini, and F. Bigler. 1998. Toxicity of the *Bacillus thuringiensis* Cry1Ab toxin on the predator *Chrysoperla carnea* (Neuroptera:Chrysopidae) using diet incorporated bioassays. *Environmental Entomology* 27:1255–1263.

4. Meier, M., and A. Hilbeck. 2001. Influence of transgenic *Bacillus thuringiensis* corn-fed prey on prey preference of immature *Chrysoperla carnea* (Neuroptera: Chrysopidae). *Basic and Applied Ecology* 2:35–44.

5. Venkateswerlu, G., and G. Stotzky. 1992. Binding of the protoxin and toxin proteins of *Bacillus thuringiensis* subsp. *kurstaki* on clay minerals. *Current Microbiology* 25:1–9.

6. Tapp, H., and G. Stotzky. 1995. Insecticidal activity of the toxins from *Bacillus thuringiensis* subspecies *kurstaki* and *tenebrio* adsorbed and bound on pure and soil clays. *Applied Environmental Microbiology* 61:1786–1790.

7. Tapp, H., and G. Stotzky. 1998. Persistence of the insecticidal toxins from *Bacillus thuringiensis* subsp. *kurstaki* in soil. *Soil Biology and Biochemistry* 30:471–476.

8. Saxena, D., S. Flores, and G. Stotzky. 1999. Insecticidal toxin from *Bacillus thuringiensis* in root exudates of transgenic corn. *Nature* 402:480.

9. Saxena, D., and G. Stotzky. 2001. *Bacillus thuringiensis* (Bt) toxin released from root exudates and biomass of Bt corn has no apparent effect on earthworms, nematodes, protozoa, bacteria, and fungi in soil. *Soil Biology and Biochemistry* 33:1225–1230.

10. Palm, C.J., D.L. Schaller, K.K. Donegan, and R.J. Seidler. 1996. Persistence in soil of transgenic plant-produced *Bacillus thuringiensis* var. *kurstaki* δ-endotoxin. *Canadian Journal of Microbiology* 42:1258–1262.

11. Sims, S.R., and L.R. Holden. 1996. Insect bioassay for determining soil degradation of *Bacillus thuringiensis* subsp. *kurstaki* CryIA(b) protein in corn tissues. *Environmental Entomology* 25:659–664.

12. Saxena, D., S. Flores, and G. Stotzky. 2002. Vertical movement in soil of insecticidal Cry1Ab protein from *Bacillus thuringiensis*. *Soil Biology and Biochemistry* 34:111–120.

13. Candolfi, M.P., K. Brown, C. Grimm, B. Reber, and H. Schmidli. 2004. A faunistic approach to assess potential side-effects of genetically modified Bt-corn on non-target arthropods under field conditions. *Biocontrol Science and Technology* 14:129–170.

14. Head, G., B. Freeman, W. Moar, J. Ruberso, and S. Turnipseed. 2001. Natural enemy abundance in commercial Bollgard and conventional cotton fields. *Proceedings of the Beltwide Cotton Conference* 2:796–798.

15. Stewart, C.N., Jr., and S.K. Wheaton. 2001. GM crop data—agronomy and ecology in tandem. *Nature Biotechnology* 19:3.

16. Heard, M.S., et al. 2003. Weeds in fields with contrasting conventional and genetically modified herbicide-tolerant crops. I. Effects on abundance and diversity. *Philosophical Transactions of the Royal Society of London Series B* 358:1819–1832.

17. Brooks, D.R., et al. 2003. Invertebrate responses to the management of genetically modified herbicide-tolerant and conventional spring crops. I. Soil-surface-active invertebrates. *Philosophical Transactions of the Royal Society of London Series B* 358:1847–1862.

18. Haughton, A.J., et al. 2003. Invertebrate responses to the management of genetically modified herbicide-tolerant and conventional spring crops. II. Within-field epigeal and aerial arthropods. *Philosophical Transactions of the Royal Society of London Series B* 358:1863–1877.

19. Mitchell, P. 2003. Europe responds to UK's GM field trials. *Nature Biotechnology* 21:1418–1419.

20. Chassey, B., C. Carter, M. McGloughlin, A. McHughen, W. Parrott, C. Preston, R. Roush, A. Shelton, and S.H. Strauss. 2003. UK field-scale evaluations answer wrong questions. *Nature Biotechnology* 21:1429–1430.

21. Freckleton, R.P., W. J. Sutherland, and A.R. Watkinson. 2003. Deciding the future of GM crops in Europe. *Science* 302:994–996.

Chapter 8

1. Tabashnik, B.E., Y.-B. Liu, T. Malvar, D.G. Heckel, L. Masson, and J. Ferre. 1998. Insect resistance to *Bacillus thuringiensis*: uniform or diverse? *Philosophical Transactions of the Royal Society of London B* 353:1751–1756.

2. Gould, F., A. Anderson, A. Reynolds, L. Bumgarner, and W. Moar. 1995. Selection and genetic analysis of a *Heliothis virescens* (Lepidoptera: Noctuidae) strain with high levels of resistance to *Bacillus thuringiensis* toxins. *Journal of Economic Entomology* 88:1545–1559.

3. Zhao, J.-Z., J. Cao, Y. Li, H.L. Collins, R.T. Roush, E.D. Earle, and A.M. Shelton. 2003. Transgenic plants expressing two *Bacillus thuringiensis* toxins delay insect resistance evolution. *Nature Biotechnology* 21:1493–1497.

4. Stewart, C.N., Jr. 1999. Insecticidal transgenes into nature: gene flow, ecological effects, relevancy and monitoring. In *Symposium Proceedings No. 72, Gene Flow and Agriculture—Relevance for Transgenic Crops* (P.J.W. Lutman, Ed.), pp. 179–190. Proceedings of a Symposium held at the University of Keele, April 12–14, 1999. British Crop Protection Council, Surrey, U.K.

5. Pulliam, D.A., D.L. Williams, R.M. Broadway, and C.N. Stewart, Jr. 2001. Isolation and characterization of a serine proteinase inhibitor cDNA from cabbage and its antibiosis in transgenic tobacco plants. *Plant Cell Biotechnology and Molecular Biology* 2:19–32.

6. Cao J., A.M. Shelton, and E.D. Earle. 2001. Gene expression and insect resistance in transgenic broccoli containing a *Bacillus thuringiensis cry1Ab* gene with the chemically inducible PR-1a promoter. *Molecular Breeding* 8:207–216.

Chapter 9

1. Vaucheret, H., C. Beslin, and M. Fagard. 2001. Post-transcriptional gene silencing in plants. *Journal of Cell Science* 114:3083–3091.

2. Fitch, M., F.M. Manshardt, D. Gonsalves, and J.L. Slightom. 1993. Transgenic papaya plants from *Agrobacterium* mediated transformation of somatic embryos. *Plant Cell Reports* 12:245–249.

3. Ferreira, S.A., K.Y. Pitz, R. Manshardt, F. Zee, M. Fitch, and D. Gonsalves. 2002. Virus coat protein transgenic papaya provides practical control of papaya ringspot virus in Hawaii. *Plant Disease* 86:101–105.

4. Tricoli, D.M., K.J. Carney, P.E. Russell, J.R. McMaster, D.W. Groff, K.C. Hadden, P.T. Himmel, J.P. Hubbard, M.L. Boeshore, and H.D. Quemada. 1995. Field evaluation of transgenic squash containing single or multiple virus coat protein gene constructs for resistance to cucumber mosaic virus, watermelon mosaic virus 2, and zucchini yellow mosaic virus. *Bio/Technology* 13:1458–1465.

5. Powell, P.A., D.M. Stark, P.R. Sanders, and R.N. Beachy. 1989. Protection against tobacco mosaic virus in transgenic plants that express tobacco mosaic virus antisense RNA. *Proceedings of the National Academy of Sciences USA* 86:6949–6952.

6. Worobey, M., and E.C. Holmes. 1999. Evolutionary aspects of recombination in RNA viruses. *Journal of General Virology* 80:2535–2543.

7. Gal, S., B. Pisan, T. Hohn, N. Grimsley, and B. Hohn. 1992. Agroinfection of transgenic plants leads to viable cauliflower mosaic virus by intermolecular recombination. *Virology* 187:525–533.

8. Schoelz, J.E., and W.M. Wintermantel. 1993. Expansion of viral host range through complementation and recombination in transgenic plants. *Plant Cell* 5:1669–1679.

9. Borja, M., T. Rubio, H.B. Scholthof, and A.O. Jackson. 1999. Restoration of wild-type virus by double recombination of tombusvirus mutants with a host transgene. *Molecular Plant-Microbe Interactions* 12:153–162.

10. Thomas, P.E., S. Hassan, W.K. Kaniewski, E.C. Lawson, and J.C. Zalewski. 1998. A search for evidence of virus/transgene interactions in potatoes transformed with the potato leafroll virus replicase and coat protein genes. *Molecular Breeding* 4:407–417.

11. Aaziz, R., and M. Tepfer. 1999. Recombination between genomic RNAs of two cucumoviruses under conditions of minimal selection pressure. *Virology* 263:282–289.

12. Rubio, T., M. Borja, H.B. Scholthof, and A.O. Jackson. 1999. Recombination with host transgenes and effects on virus evolution: an overview and opinion. *Molecular Plant-Microbe Interactions* 12:87–92.

13. Allison, R.F., W.L. Schneider, and A.E. Greene. 1996. Recombination in plants expressing viral vectors. *Seminars in Virology* 7:417–422.

Chapter 10

1. Phuntumart, V. 2003. Transgenic plants for disease resistance. In *Transgenic Plants: Current Innovations and Future Trends* (C.N. Stewart, Jr., ed.), pp. 179–216. Horizon Scientific Press, Wymondham, U.K.

2. De la Fuente, J.M., V. Ramirez-Rodriguez, K.L. Cabrera-Ponce, and L. Herrera-Estrella. 1997. Aluminum tolerance in transgenic plants by alteration of citrate synthase. *Science* 276:1566–1568.

3. Ezaki, B., R.C. Gardner, Y. Ezaki, and H. Matsumoto. 2000. Expression of aluminum-induced genes in transgenic arabidopsis plants can ameliorate aluminum stress and/or oxidative stress. *Plant Physiology* 122:657–665.

4. Committee on Environmental Impacts associated with Commercialization of Transgenic Plants, National Research Council. 2002. *Environmental Effects of Transgenic Plants: The Scope and Adequacy of Regulation.* National Academy Press, Washington, D.C.

5. Halfhill, M.D., H.A. Richards, S.A. Mabon, and C.N. Stewart, Jr. 2001. Expression of GFP and Bt transgenes in *Brassica napus* and hybridization and introgression with *Brassica rapa*. *Theoretical and Applied Genetics* 103:659–667.

6. Warwick, S.I., M.J. Simard, A. Légère, L. Braun, H.J. Beckie, P. Mason, B. Zhu, and C.N. Stewart, Jr. 2003. Hybridization between *Brassica napus* L. and its wild relatives: *B. rapa* L., *Raphanus raphanistrum* L. and *Sinapis arvensis* L., and *Erucastrum gallicum* (Willd.) O.E. Schulz. *Theoretical and Applied Genetics*.

7. Halfhill, M.D., R.J. Millwood, A.K. Weissinger, S.I. Warwick, and C.N. Stewart, Jr. 2003. Additive transgene expression and genetic introgression in multiple GFP transgenic crop × weed hybrid generations. *Theoretical and Applied Genetics* 107:1533–1540.

8. Stewart, C.N., Jr., M.D. Halfhill, P.L. Raymer, and S.I. Warwick. 2003. Transgene introgression and consequences in *Brassica*. In *Conference Proceedings: Introgression from Genetically Modified Plants into Wild Relatives and its Consequences* (H. Den Nijs, D. Bartsch, and J. Sweet, eds.). Amsterdam, January 22–24, 2003.

9. Adam, D. 2003. Transgenic crop trial's gene flow turns weeds into wimps. *Nature* 421:462.

10. Stewart, C.N., Jr., M.D. Halfhill, and S.I. Warwick. 2003. Transgene introgression from genetically modified crops to their wild relatives. *Nature Reviews Genetics* 4:806–817.

Chapter 11

1. Shelton, A.M., J.-Z. Zhao, and R.T. Roush. 2002. Economic, ecological, food safety, and social consequences of the deployment of Bt transgenic plants. *Annual Review of Entomology* 47:845–881.

2. Ehrlich, P.R. 1968. *The Population Bomb*. Ballantine Books, New York.

3. Green Spirit Website. Available: http://www.greenspirit.com

4. Ecological Society of America Website. Available: http://www.esa.org/pao/statements/esagmo.htm.

5. Rissler, J., and M. Mellon. 1996. *The Ecological Risks of Engineered Crops*. MIT Press, Cambridge, Mass.

6. Goklany, I.-M. 1998. Saving habitat and conserving biodiversity on a crowded planet. *Bioscience* 48:941–953.

7. Trewavas, A.J. 2001. The population/biodiversity paradox. Agricultural efficiency to save wilderness. *Plant Physiology* 125:174–179.

8. Maeder, P., A. Fliessbach, D. Dubois, L. Gunst, P. Fried, and U. Niggli. 2002. Soil fertility and biodiversity in organic farming. *Science* 296:1694–1697.

Chapter 12

1. Kooshki, M., A. Mentewab, and C.N. Stewart, Jr. 2003. Pathogen inducible reporting in transgenic tobacco using a GFP construct. *Plant Science* 165:213–219.

2. Stewart, C.N., Jr. 2001. The utility of green fluorescent protein in transgenic plants. *Plant Cell Reports* 20:376–382.

3. Mentewab, A., V. Cardoza, and C.N. Stewart, Jr. 2003. Promoter analysis of *Arabidopsis* genes differentially expressed upon exposure to trinitrotoluene. Abstract 49. *Plant Genetics 2003 Abstract Book*. Snowbird, Ut., October 22–26, 2003.

4. Hannink, N., S.J. Rosser, C.E. French, A. Basran, J.A.H. Murray, S. Nicklin, and N.C. Bruce. 2001. Phytodetoxification of TNT by transgenic plants expressing a bacterial nitroreductase. *Nature Biotechnology* 19:1168–1172.

5. Pilon-Smits, E.A., S. Hwang, C.M. Lytle, Y. Zhu, J.C. Tai, R.C. Bravo, and Y. Chen. 1999. Overexpression of ATP sulfurylase in Indian mustard leads to increased selenate uptake, reduction, and tolerance. *Plant Physiology* 119:123–132.

6. Rugh, C.L., D. Wilde, N.M. Stack, D.M. Thompson, A.O. Summers, and R.B. Meagher. 1996. Mercuric ion reduction and resistance in transgenic *Arabidopsis thaliana* plants expressing a modified bacterial *merA* gene. *Proceedings of the National Academy of Sciences USA* 93:3182–3187.

7. Bizily, S., C.L. Rugh, A.O. Summers, and R.B. Meagher. 1999. Phytoremediation of methylmercury pollution: *merB* expression in *Arabidopsis thaliana* confers resistance to organomercurials. *Proceedings of the National Academy of Sciences USA* 96:6608–6813.

8. Rugh, C.L., J.F. Senecoff, R.B. Meagher, and S.A. Merkle. 1998. Development of a transgenic yellow poplar for mercury phytoremediation. *Nature Biotechnology* 16:925–928.

9. Krämer, U., and A.N. Chardonnens. 2001. The use of transgenic plants in the bioremediation of soils contaminated with trace elements. *Applied Microbiology and Biotechnology* 55:661–672.

10. Jouanin, L., T. Goujon, V. de Nadaï, M.-T. Martin, I. Mila, C. Vallet, B. Pollet, A. Yoshinaga, B. Chabbert, M. Petit-Conil, and C. Lapierre. 2000. Lignification in transgenic poplars with extremely reduced caffeic acid O-methyltransferase activity. *Plant Physiology* 123:1363–1374.

11. Brinch-Pedersen, H., L.D. Sørensen, and P.B. Holm. 2002. Engineering crop plants: getting a handle on phosphate. *Trends in Plant Science* 7:118–125.

12. Verwoerd, T.C., P.A. van Paridon, A.J.J. van Ooyen, J.W.M. van Lent, A. Hoekema, and J. Pen. 1995. Stable accumulation of *Aspergillus niger* phytase in transgenic tobacco leaves. *Plant Physiology* 109:1199–1205.

13. Denbow, D.M, E.A. Grabau, G.H. Lacy, E.T. Kornegay, D.R. Russell, and P.F. Umbreck. 1998. Soybeans transformed with a fungal phytase gene improve phosphorus availability for broilers. *Poultry Science* 77:878–881.

14. Poirier, Y., D.E. Dennis, K. Klomparens, and C. Somerville. 1992. Polyhydroxybutyrate, a biodegradable thermoplastic, produced in transgenic plants. *Science* 256:520–523.

15. Slater, S., T.A. Mitsky, K.L. Houmiel, M. Hao, S.E. Reiser, N.B. Taylor, M. Tran, H.E. Valentin, D.J. Rodriguez, D.A. Stone, S.R. Padgette, G. Kishore, and K.J. Gruys. 1999. Metabolic engineering of *Arabidopsis* and *Brassica* for poly(3-hydroxybutyrate-co-3-hydroxyvalerate) copolymer production. *Nature Biotechnology* 17:1011–1016.

Chapter 13

1. Kooshki, M., A. Mentewab, and C.N. Stewart, Jr. 2003. Pathogen inducible reporting in transgenic tobacco using a GFP construct. *Plant Science* 165:213–219.

2. Robbins, J. 2002. Farms and growth threaten a sea and its creatures. *New York Times*. April 2, 2002, late edition, section F, p. 3.

3. Pimentel, D.S., and P.H. Raven. (2000). Bt corn pollen impacts on nontarget Lepidoptera: assessment of effects in nature. *Proceedings of the National Academy of Sciences USA* 97:8198–8199.

4. Strauss, S.H. 2003. Genomics, genetic engineering, and domestication of crops. *Science* 300:61–62.

Index